Human Power

The rapid and tumultuous technological transformation of our societies has triggered a self-exploratory public debate about what it means to be human. What are our human potential, talents, and powers – what. essentially. is our place in the modern world?

Will a culture of machines out-compete a culture of humanity?

In this thought-provoking but ultimately hopeful book, scholar and leading technology critic Gry Hasselbalch invites readers to reflect on the shifting dynamics between humans and the AI-powered technologies increasingly shaping our world. Exploring the distinctiveness of human power, the book addresses current debates about technology that portray humans as powerless and flawed – in essence outdated software in dire need of a technological fix – arguing that human power must remain central in discussions about AI and technology. It investigates seven key traits which set humans apart from machines: Creativity, Intuition, Emotion, Life, Defiance, Love, and Wisdom. Drawing on interviews and examples from across arts – including literature, visual arts, film and music – and from technology politics and policymaking, *Human Power* explains how these traits provide a foundation for a new politics in the AI Machine Age. One that does not diminish and reduce human power, but actively protects and reinforces it.

If human power is not a computational process, then what is it? ***Human Power: Seven Traits for the Politics of the AI Machine Age*** gives human power – our humanity and fundamental "humanness" – a renewed voice in a debate dominated by fears and preconceptions about technological power. This important new work will appeal to journalists, policymakers, artists, and educators—and anyone else interested in the rapidly growing role of AI and digital technology in our lives.

Gry Hasselbalch, PhD

Human Power
Seven Traits for the Politics
of the AI Machine Age

Gry Hasselbalch

CRC Press
Taylor & Francis Group
Boca Raton London New York

CRC Press is an imprint of the
Taylor & Francis Group, an **informa** business

Designed cover image: "Yggdrasil", 2015, by Ida Kvetny

First edition published 2025
by CRC Press
2385 NW Executive Center Drive, Suite 320, Boca Raton FL 33431

and by CRC Press
4 Park Square, Milton Park, Abingdon, Oxon, OX14 4RN

CRC Press is an imprint of Taylor & Francis Group, LLC

ISBN: 978-1-032-86513-3 (hbk)
ISBN: 978-1-032-69816-8 (pbk)
ISBN: 978-1-003-52785-5 (ebk)

DOI: 10.1201/9781003527855

Typeset in Minion
by KnowledgeWorks Global Ltd.

*To my human power Clara Astrid. The most unique
human being with the kind of human potential
that can never be restrained or repressed.*

Power to you, my love.

Contents

Introduction

Humanity in the AI Machine Age

Sometimes, we don't use our imagination. It happens, really, because we tend only to go so far as our human eyes can see, the body feel, and thoughts can grasp. In this book, however, I ask you to use your imaginative powers and revel in humanity's great potential, especially in these seven traits of human power: creativity, intuition, emotion, life, defiance, love, and wisdom. You will see that the power of human life is majestic and luminous, intuitive, dynamic, thrilled, and unpredictable; it moves where it is restrained, imposes arbitrariness where perfect predictions rule, and makes noise where it is supposed to be silent – and we urgently need a technology politics devoted to a power like that. Why do we need this now? I'll tell you why.

In the history of computing, although leading to incredible inventions and significant social and economic advancements such as the World Wide Web, big data, and artificial intelligence, I believe that we didn't use our imagination well. From the beginning, we developed, adopted, and deployed computer-based technologies based on limited ideas about what it means to be human. Human power – like our intelligence, creativity, intuition, feeling, and compassion – were, and are still today, envisioned as traits that machines can imitate, match, extend in frictionless patterns, and even surpass. What is even more worrying is that this simpleminded technological solutionism, imitating the weakest and most simple version of humanity and human power, has been, in fact, also sieving into our

DOI: 10.1201/9781003527855-1

political imagination. Of course, in particular, global technology politics has been busy hasting after the computational transformation of the world. It had to, and still, it was always one step behind, always trying to catch up. At first, we were predominantly occupied with the technical and legal efficiency and infrastructural interoperability of what we were told would be the direct extension of our human communities, a glorious "global information village". Then, we had to swiftly scramble to create technology policies in defence against the consequences happening in a flash of limited views of what it means to be human – against weak privacy protection, unbounded digital surveillance, reductive technological democracy, and discriminatory algorithms replacing human decision-making. We were compelled to do so due to the significant social and ethical ramifications now coming to light.

Today, with the rocket launch of generative artificial intelligence into every sector and every sphere of our lives, we have even started questioning our human potential. If a machine can write a perfectly sounding song, scientific paper, make a movie or create a piece of art, what is it then that humanity can do? What is so unique about human beings? Whether we should worry about this or not – and I will try to provide some answers to these questions in this book – the fear is very real, and of course, humans act on emotions such as fear. We must acknowledge that. It is only human to do so. All the same, it is time now to regain our self-confidence and use our imagination better because if you genuinely think about it, human power can take us further than that.

The seven traits of human power I present in this book are distinct traits that I will show you a machine cannot replace. At the same time, they are the human traits most challenged and weakened by technologies developed to imitate human power. Of course, there is no doubt that digital technologies also reinforce human power. Digital health technology provides precision and personalised medicine, allows us to monitor our health, and receive remote medical care, which can extend our lives. The internet helps us to expand our knowledge and communication globally. Similarly, virtual reality technologies allow us to explore new places and learn new skills that we might not have access to otherwise. Overall, technology empowers us by expanding our capabilities.

Yet, at the same time, the kind of human power I present in this book is fundamentally challenged and warped in the life of technological dominance that we are living today. Simplistic automated systems are sieving through data to grade children in schools, distribute family benefits, loans, and medical care, identify individual life risk factors, and influence democratic elections. Human desires and interests are identified, we are categorised, and general trends are datafied, analysed, and materialised in the films, TV series, and music scores of our lives' most precious moments. Today, all that is most human is dissected and distilled in digital data systems, often with devasting human consequences.

In the United States, an AI system was used to score the health status of patients waiting for a kidney transplant, assigning healthier scores to African American people and thus affecting decisions regarding their eligibility for a transplant.[1] In the United Kingdom, a national exam board deployed an algorithm to generate student grades, which overruled teacher assessments of each student. This resulted in the grades of students from large state schools decreasing while those attending smaller fee-paying schools increased.[2] In the Netherlands, a national digital welfare fraud detection system was deployed primarily in low-income neighbourhoods to detect the likelihood of an individual committing tax or benefit fraud.[3] And in China, jaywalking, not paying a fine, and other forms of public misbehaviour are added to a local social credit scoring system that assesses and provides citizens with a public score.[4]

Digital systems and registers of power exist everywhere in all sectors and spheres of our lives. Even so, human power stands out in this mesh of technologically shaped lives and societies. It is an unpredictable, disorderly, and self-reflective power that does not fit easily into the classification systems of data technologies nor in the forebodings of their predictions, escaping their assessments and confusing their decisions. When students in the United Kingdom received automatically generated grades, outraged protests broke out all over the country, and images of students with different messages travelled around the world: "Students. Not stats!", "Trust our teachers!" Human power can be magnificent like that – critical, forceful, full of emotion, and projective of experience in time and space. And we need this human power as the foundation for the politics and governance of the big data and AI age. We need to see beyond just economic progress and technological perfection and do our best to preserve, boost, reinforce, and support these traits of human power. Not diminishing and reducing them with simplistic machines. If we do this, we will not only ensure our rights and well-being as individual human beings, but we can think of innovation in a different way that will enable the design of alternative socio-technical infrastructures, enhancing human power rather than replacing it.

WHY

Before you continue reading, I want to share a few things about the book – why I wrote it and provide you with some explanations for what you might consider its weaknesses.

In 2021, amid a lingering pandemic and the initial phases of the negotiation of the world's first AI law in the European Union, I was, at once, taken aback and unsettled by an abrupt change of direction in the public debate about technology. I had been working with internet, digital and data policies, and public awareness for almost two decades – the last couple of years with AI. In the European Union's high-level group on AI, we had created ethical principles for AI that were now being transformed into law. Around the world, similar initiatives had been emerging for some years. Global actors were increasingly taking an "ethical stand" on emerging technologies – with catchphrases such as "trustworthy" or "human-centric" AI. We were getting somewhere – also in the public debate that was finally starting to really question the largely unchecked power of the big tech industry. But then suddenly, it seemed that the rapid and tumultuous technological transformation of society, that the pandemic had also accelerated, was triggering a renewed self-exploratory public debate – about what it means to be human, our potential, powers, and place in the modern world, and this only intensified in 2022 when new generative AI models were spiralled into the public sphere without warning. It felt as if we were starting all over. As if all the humanistic ideals of past generations were suddenly up for debate. Are humans just "outdated software"[5] in dire need of a technological fix? Will a culture of machines out-compete a culture of humanity? The answers didn't leave much certainty about the future of humanity. They were feeble, insecure, and, at best, dogmatic. Promoting moral abstract principles without really any grounded reference to the vast expressions of human power that human history has produced.

I started writing this book to give evidence to human power and the humanistic perspective in a public debate dominated by banters on technological power. If human power is not a computational process, easily deduced, reduced, represented, and discerned, like the AI evangelists, for instance, are telling us, then what is it? And could we imagine a more nuanced and evidenced exploration of human power as a foundation for a new technology politics in the age of big data and AI? A new technology *politics* in the most basic sense of the word, that is, the agreements we humans make in our various groups and communities. Could we think of human power as not only the basis of the technology politics of lawmakers and governments but also as the foundation of the agreements and promises we make in companies, classrooms, in the science lab and at home in the family about what role technology should play in our cultures, societies, and individual lives? What if we made these seven traits of human power – creativity, intuition, emotion, life, defiance, love, and wisdom – into the guiding directive of a *politics for everyday life* in the AI machine age in general? Would we change our priorities? Would we feel more empowered to challenge unfounded claims about technological powers? Would we do things differently?

Now, you will see that my voice is very visible in the book and thus it carries a lot of my Scandinavian (Danish), European, and especially Western cultural background. I wanted to make this very clear from the beginning, because this also means that although I am addressing the fundamental "humanness" of being human and the technology politics, we urgently need to empower and protect humanity; I do illustrate this argument with what I know best and, therefore, also do not always manage to bring in the richness of the world's cultural perspectives and examples that I would have liked to. Please accept my unreserved apologies for that.

Also, I cannot stress enough that I am not presenting a new definition of human power. What I do is to combine existing ideas and cultural expressions to introduce a technology politics with a humanistic perspective. Therefore, when describing the seven traits of human power, my intention was never to develop exhaustive definitions, it was to give these traits a voice in a public debate about the role of technology in a human world. This is also why the exploration of the traits is limited to that context.

Finally, I admit that I am quite rudimentary in the book when I use the term "technology". In fact, technology, which stems from the Greek *tekhnē*, just means the application of science and knowledge, our skills and techniques developed and used by humans to create things. Most political activities are, in some way or another, tackling the technologies used and developed by humans. Yet, in the early 21st century, there has been a specific focus on national, regional, and international politics for the technologies of this age – digital technologies, big data, AI, and the development of what is referred to as the Information Society and now increasingly the AI machine age.

We have thus seen in the last couple of decades international political agreements on the technical and economic development, as well as legal and technical interoperability of these technologies, and increasingly also on protecting individual human rights, liberties, and well-being and ensuring just power balances in democratic societies. Nevertheless, I believe we need an even broader perspective to preserve the power of humanity altogether. One crucial component of a standpoint as such is indeed individual liberty and empowerment; another is the preservation of democratic societies. But furthermore, there is an urgent need for a more conscious effort to protect and enhance the human power we share as human beings, those human traits and conditions that bind us together and push us forward to do amazing things, make sense of the world through art and poetry, revolt and change the state of affairs.

We need this humanistic approach in the official policies of our governments and intergovernmental institutions. We also urgently need it in the politics of the public debate, among journalists, the designers of technology services and products, scholars, and industry leaders, and we need it ourselves in our everyday lives. We must regain confidence in our human power in all spheres of human life and actively work towards defending, preserving, and boosting it.

Consider that all the human beings exposed to an unfair digital data system of power, although their individual experiences differ, share a sense of disempowerment. This collective experience of human disempowerment should be at the heart of the concerns of any technology politics today. So, if we genuinely want to preserve human power in the AI machine age, we need to look away from simple technologies that imitate human power and instead focus on actual human power.

THE BOOK

The book consists of seven core chapters and a part one and two, each tackling human power in a different way.

In the book's first part, I explore different views on the conditions and nature of human power and how these affect our understanding, and not the least confidence in, or worries about, our human potential and limitations. Human power has, throughout time, been challenged by different conceptions of fate and destiny, from metaphysical justifications based on the reasoning of gods to those grounded in the laws of physics and the promise of progress and objective science. Today, a new challenge takes form in the promises of technology viewed as an end goal in and by itself. "Technology" has become the projection of human fate, just like the superstition of ancient times. Increasingly, we imagine the shape and direction of our human future through the binoculars of technology, and we seem to trust in its potential to empower and guide humankind without question. To counter this narrative, I in this part of the book, trace the humanistic tradition in some of the most rebellious cultural and social trends of the 20th century. I also introduce the general components of human power, such as love, brutality, memory, condition, nature, and collective power, with reference to critical theory and examples from literature, human stories, art, films, and music. The key guiding question of this part of the book is: are we overlooking human power's vast potential when fixating on its darker side?

In Chapters 1–7, I explore seven traits of human power. These traits are presented through narratives that incorporate human stories, art, film, music, and literature and with reference to theories tackling each trait from different perspectives. I aim to illustrate the uniquely "human character" of these traits. At the same time, I want the reader to consider these human traits in the context of technological power, why each chapter also poses questions regarding the power of technologies that imitate human power, such as Artificial intelligence. How do they differ? How does technological power challenge these human traits? Where does our current perception of technological power fall short?

In the book's part two, I briefly outline the development of global technology politics in the 21st century in three phases: one for technology, one for ethics and society, and one for humanity. In history, they are intertwined, and none stands alone; mainly, they extend into one another. However, looking back at the history of early 21st-century technology politics, the three phases represent each of their prevailing political calls for action. While the first two did achieve the development of globally shared values-based principles, emphasising global interoperability, openness, human rights, democracy, and specifically the "human-centric" approach, they did not manage to fully translate all these principles into practice. Nevertheless, in the third wave of technology politics, we see a much stronger emphasis on the unique qualities of humanity emerging. This is, therefore, also the phase that I, in this last part of the book, aim to present by tracing it back and pointing forward in technology politics with these six themes: human rights, the human-centric approach, emotional politics, a new social contract, technology diplomacy, and the global approach. In this part of the book, I also include in more detail the views of people working in concrete and influential ways in 21st-century technology politics.

NOTES

1. Simonite, T. (2020, October 26th) "How an algorithm blocked kidney transplants to black patients", *Wired*, https://www.wired.com/story/how-algorithm-blocked-kidney-transplants-black-patients/
2. Hern, A. (2020, August 14th) "Do the maths: why England's A-level grading system is unfair", *The Guardian*, https://www.theguardian.com/education/2020/aug/14/dothe-maths-why-englands-a-level-grading-system-is-unfair; Hern, A. (2020, August 21st) "Ofqual's A-level algorithm: why did it fail to make the grade?", *The Guardian*, https://www.theguardian.com/education/2020/aug/21/ofqualexams-algorithm-why-did-it-fail-make-grade-a-levels
3. Heikkila, M. (2022, March 22nd) "Dutch scandal serves as a warning for Europe over risks of using algorithms", *Politico*, https://www.politico.eu/article/dutch-scandal-serves-as-a-warning-for-europe-over-risks-of-using-algorithms/
4. Yang, Z. (2022, November 22nd) "China just announced a new social credit law. Here's what it means", *MIT Technology Review*, https://www.technologyreview.com/2022/11/22/1063605/china-announced-a-new-social-credit-law-what-does-it-mean/
5. Lunau, K. (2013, October 14th) "Google's Ray Kurzweil on the quest to live forever", *Maclean's*, https://www.macleans.ca/society/life/how-nanobots-will-help-the-immune-system-and-why-well-be-much-smarter-thanks-to-machines-2/

I

Human Power

What Machines Don't Have

It is the 5th of June 1989. A man in black slacks, a white shirt, and what looks like shopping bags in each stretched-out arm stands immobile in front of a string of big military tanks crawling slowly towards him. Mammoth moving chunks of high hardness steel in the form of "state of the art" military technology steadily approaching a tiny human figure. The man doesn't move, his shoulders slightly crouched forward. Then, the tanks stop just in front of him, and in a sudden burst of energy, the man lashes out with one arm at the tanks. Quickly and forcefully, just one arm and a shopping bag sway in and then out in a "go away" (angry, relieved, sad?) movement. His feet and body don't move, just that one arm. Never does he move out of position to shy away from the tanks.

How small and yet massively powerful the lives and acts of human beings can be. "Tank Man", who stood in front of the tanks on Tiananmen Square in Beijing in 1989, the day after the 4th of June massacre of Chinese student protesters, is a reminder of the human power of resistance. There are countless examples like these in history – powerful acts projected by human life, emotion, solidarity, love, and others of these traits of human power. We will never know precisely the human emotion, intuition, reflection, and memory that planted Tank Man's feet solidly on the ground before the tanks that day. Still, we can imagine it was one of these human traits.

DOI: 10.1201/9781003527855-2

What do stories like this tell us about human power? That human power is fierce and has been a driver of action and change throughout history. We also understand that the human power that seems to be competing and juxtaposed with the technological power that we are also exploring in this book is most powerful precisely because it has something that technology can represent but never truly replace: it is unpredictable, lived, emotional, it loves, and it is painfully aware of its impermanence and its unique position in time and space.

Let's take a step back and look at the technological developments not only in military tanks but in all spheres of our human existence over the last century. The most stunning factor is the acceleration of the technologisation of the world. Every aspect of human life, society, and even the natural environment was transformed and configured by human technology so fast and with such power that it is almost impossible to grasp with our unique but also very slow human bodies and minds. From writing down books by hand to printing hundreds using the first movable metal types, and then a little more than 1 billion books each year with modern printing presses. From travelling by horse and boat to moving across the land by locomotive to flying across the sky in aeroplanes. From living at the mercy of nature, the Gods of heathens – the sun, thunder, soil, and water – to invading the particles of nature, leaving our carbon footprints. From drawing the earth on parchment maps and metal globes to creating a panoramic view of the world by capturing digital imagery with special cameras that simultaneously collect images in multiple directions and stitch them into a single 360° image map. From collecting and storing digital data in huge, senseless big data repositories to sieving through all that data, making sense of it all, and generating conversational language models or predicting the development of a virus based on these data.

Just 100 years ago, a man like Tank Man, standing in front of the technological excellence of one regime of power during a mass protest, would have had a different name. Most probably, he would have been "Horse Man", placing his human body firmly in front of a line of horse-riding police officers with sticks and perhaps guns. At the end of the 19th century, police horses provided a "height advantage" for police officers during mass gatherings.[1] Most often viewed and treated as machines rather than animals, police horses were trained and cultivated for the specific purpose of controlling crowds; their resilience was tested, for example, by throwing firecrackers underneath them.[2]

Today, Tank Man might not have been. Or it would be not easy to find a name for him. "Multi-shot Riot Man"? "Chemical Irritant Man"? "Surveillance Camera Man"? "AI Man"? Indeed, his circumstances would have been very different. "Crowd control technologies" have evolved. Countries worldwide are deploying various technologies to monitor, control, and predict the movements and actions of gatherings of people. Kinetic impact weapons, multi-shot riot guns, marker dye and chemical-irritants, tear gas projectiles, water cannons firing bullets of water, 50,000-V riot shields, and side-handle batons.[3] In a very short period, companies and countries developing and supplying crowd control technologies have increased immensely. Crowd control technology policies and legislation have been adopted, and international agreements have been made. And city infrastructures have been adapted into socio-technical infrastructures fit for social control.

The smart city is not only an intelligent design that supports citizens in their various doings, from finding a parking spot or the nearest pizza restaurant. The centralised City Brain artificial intelligence (AI) data system developed by the private Chinese tech giant Alibaba adopted in cities across China monitors video footage of traffic, looking out for signs of collisions or accidents to alert the police. It combines data from the transportation bureau, public transportation systems, a mapping app, and hundreds of thousands of cameras. Road accidents are, for example, automatically detected so they can be responded to faster, and illegal parking is tracked live. The system is used by Chinese law enforcement and the Chinese government to control the cities.[4]

We would most certainly know who he was had Tank Man walked in front of the tanks today. AI technologies that can identify a face by matching it with a database of faces, so-called facial recognition technologies, are deployed in many smart city infrastructures worldwide. Facial recognition is not the only way a person can be identified. All sorts of "biometric identification" systems are currently deployed or tested in public spaces "iris recognition", "gait recognition" (systems that discern age and gender based on the way people walk), "voice recognition" combined with "gun-shot detection" (used in city zones by for example US police).[5]

We might even be able to detect Tank Man's emotions, and he would most probably be stopped before walking in front of the tanks. AI technologies have also been developed to predict people's emotions by scanning the data on their faces and interpreting micro-movements in the face and body language.[6] It's considered useful when controlling crowds of people gathering and pre-empting what is considered unsafe or erroneous human action during protests or any other gathering of many people, like concerts or sports events. It is also increasingly developed and adapted for other, more mundane purposes, such as employment and recruitment. A job candidate will have their face and body scanned while, for example, watching a movie, and what is often referred to in these contexts as "non-verbal cues" will be observed. Based on different estimations, a report on the emotional suitability for the position outlining various human traits, such as ability to cope with stress, social attitude, engagement, passion, and honesty, will be produced for the recruiter to use for their decision.[7]

Advanced AI systems are created to predict if an unemployed person is likely to stay unemployed or get a job in public administration. This also happens in the judicial system when judges decide whether a person will get parole or remain in prison based on an algorithm that predicts recidivism.

Unfortunately, many of these systems are reinforcing existing biases in society. Not just those in the judicial systems that decide that white men and women are less of a threat to society than black men and women.[8] The Dutch tax authorities, for example, once introduced an algorithm-based decision-making system to generate risk profiles of individuals who applied for childcare benefits. This system aimed to detect and prevent inaccurate or fraudulent applications early. One of the factors considered in the risk assessment was nationality. As a result, a significant number of parents and caregivers from predominantly low-income families were mistakenly accused of fraud by the tax authorities, with those from ethnic minority groups being disproportionately affected.[9]

Technologies used by police forces worldwide to control or tame crowds of humans are not the only type of social control technologies we are living with today, and neither are they the most common. Human life is represented, contained, restrained, and shaped within all these digital devices and systems. When we interact with public services, look up information, are entertained, do our shopping, go to the doctor, go to school or work. We are immersed in digital social control systems. However, they are a lot less obvious, more invisible, and subtle part of our human lives than the gag and the gun. We live in a society where the integration of what we could call "destiny machines"[10] is accelerating rapidly (see also the following section on the "Society of the Destiny Machine"). The fact is that all these technologies of power, because that is what they are (whether the power of police, companies, or groups of people), are conditioning human power, even restraining it.

Let's begin by exploring human power in more general terms – the conditions and nature of human power – as well as different human depictions and expressions of human power and how these affect our understanding, and not the least confidence in, or worries about, our human potential and limitations.

WHAT IS HUMAN POWER?

In the 2020s, facial recognition technology stirred controversy, prompting calls for regulation by policymakers and industry worldwide. The European Union's Artificial Intelligence Act was adopted with an in-principle ban on live mass, facial recognition, and other public biometric surveillance by police. Major tech companies like IBM, Amazon, and Microsoft stopped selling facial recognition tools to governments due to civil liberties risks.[11] In the United States, some states at one point considered banning the technology's use by local police, and federal lawmakers proposed a ban on federal agency use for surveillance.[12] Additionally, 170 civil society organisations called for a global ban on biometric recognition technologies, including facial recognition, due to surveillance concerns.[13]

The calls for banning facial recognition technology by law were forceful. Yet in all the official calls by organisations and lawmakers, the plead from one American black man, Robert Williams, after his encounter with a facial recognition system used by Detroit police stands out.

One day, when Mr Williams arrived at his house, police officers rushed in and blocked his car, handcuffed him in front of his wife and young daughters and brought him to the police station where he was held over the night. However, Mr Williams had done nothing wrong. He was arrested due to a wrongful match made by a facial recognition system between him and a thief with the same skin colour as his.

Imagine the human suffering and fear. Consider the combined feelings of love and anger when dealing with his two young daughters' impressions and emotions following the event. As he later wrote in his call for a facial recognition systems ban:

> My daughters can't unsee me being handcuffed and put into a police car. They continue to suffer that trauma (...) When my daughters encounter the coverage about what happened to me, they are reduced to tears by their memory of those awful days (...) I get angry when I hear companies, politicians, and police talk about how this technology isn't dangerous or flawed (...) If any of that was true, I wouldn't have been arrested.[14]

Mr Williams' very human plea to prohibit the use of this technology was entangled with his love for his children and wife, and it was compelled by the collective experience of the discrimination he felt as a black man in a society where wrongful arrests and unjust treatment by people that looked like himself was not the first. Nor the last. There were at least two more instances of erroneous mistakes made by facial recognition systems in the United States, leading to arrests in the years that followed Mr Williams' arrest.[15]

We see human power expressed in human pleas, experiences, and reactions to unjust conditions like these. Sensations of humanity, panic, angst of mortality, emotion and love, anger, and feelings of injustice. In decisive moments in past and current history, this type of human power has taken precedence, sometimes changing history, very often lived and articulated by human individuals and groups that have functioned as key drivers of change. With a very intense, unique power – one that suffers, is filled with emotion, dedication, intense reflection, and unpredictable and inexplicable actions. And if we are genuinely looking for it, it is in these human stories and expressions of human power we will find the answers to the one pertinent question of our time: what is human power?

Human power is love. Human power is brutality. Human power has a human memory. Human power has a human condition. Human power is a force of nature. Human power is like music. Let me explain:

Human power is love. Humans act and are propelled forward by feelings of love and connection with other human beings that are often stronger and bigger than life itself. In Hans Christian Andersen's fairy tale, "The Story of a Mother", a mother is grieving her sick and dying child when Death, disguised as an old man, comes to take the child away. The mother pleads with Death to spare her child, but Death takes the child anyway. The mother then journeys across the sea and land to find her child. "Oh, what would I not give to reach my child!" she cries, exchanging her eyes and her long black hair on the way to find her child, and finally when meeting Death once again, who presents to her the potential destiny of suffering of her child also accepts her pain of losing her child for good.

Humans will all experience and feel the power of life and love for their children or other human beings at some point in their lives. The force of human love can drive us to do many things: walk across land and sea, sacrifice our well-being, put ourselves in danger for another human being, and reach unimaginable and unattainable goals.

When the Taliban arrived in 2021 in Afghanistan 20 years after their expulsion by US troops, Assistant Director-General for the Social and Human Sciences of UNESCO Gabriela Ramos[16] was told by her staff that the 25-year-old marathonist they had sent to Afghanistan to promote sports for gender equality was stranded. There was no way to get her out. All the planes were already packed and leaving. Ramos saw the last US plane taking off, and she cried. She was told that there was nothing she could do. But then she looked at a picture of the woman's passport and thought of her own daughter, who was the same age. "How could I not help her? (…)", Ramos asked me when I spoke with her in 2023, "… I got her out, for God's sake. I got her out", she exhaled.

Sensibility and compassion for others Ramos believes you find in people everywhere. There are always people like that who, in crucial moments, will attain the unachievable when driven by these human sentiments. She often sees them in her work in UNESCO, an intergovernmental organisation set up at the end of the Second World War to rebuild the education systems of 44 countries and re-establish the "intellectual and moral solidarity of mankind".[17] The organisation with the "ethics mandate", as Ramos referred to it, today counts 193 member states and 12 associated members worldwide. She was one of the driving forces behind the recommendation on AI ethics adopted by the Member States at UNESCO's General Conference on 23rd November 2021. Usually, similar UNESCO instruments take 4–10 years to develop and get member states to agree on. However, the AI ethics recommendation took only two years due to a sense of urgency shared by all partners involved, Ramos explained:

> The digital transformation has been with us for more than three decades, and it has now become very powerful and extensive. We have realised that this is not just a question of new technological appliances; it is about building the kind of world we live in. It is a question about the future of our societies. The world is highly divided by inequalities of income and opportunities, racism, discrimination, and sexism. These problems have always been there, but they are being exacerbated by the digital transformation.

It is not unusual to see an upsurge of compassionate human power to rapidly create good governance instruments, such as the AI ethics recommendation in moments of crisis or as a response to the erosion of power. The intergovernmental organisation UNESCO was created exactly like that after the Second World War to prevent the future atrocities of human totalitarian powers. Human power is magnificent, but as the history that an organisation like UNESCO is built on tells us, it can also be violent and oppressive.

Human power is brutality. While compassion and sensibility are traits of human power, this is not what immediately springs to mind when contemplating human power. Even if you are not directly affected by the sound of bombs or the fear of persecution, you are still bombarded daily with streams of images of the result of power struggles – war, poverty, and destruction. The starving child on the ground, thin limps curled like branches of a tree, the bombs' blitz in the sky over a city, the corps on the ground, the barking commands of a uniformed officer, the guns or the gags, the exhausted factory worker. Our image of human power as an oppressive power of dictatorships and totalitarian states or just a ruthless world economy is not only real and experienced, but it also has an image and a contour made of films, newsreels, and narratives from the present and past.[18]

If you grew up in Europe in the second half of the 20th century, you carry with you the stories and mental images of the Second World War passed down from your parents and grandparents. I have one memory image like that. In the late morning, on 9th April 1940, my grandmother Tove was on tram line 1, which used to run across Copenhagen. That's when she discovered Denmark had been invaded and was now occupied. A few hours earlier, German soldiers had entered from the coastline at the Copenhagen harbour of Langelinje. They were now in the streets, accompanied by the noise of military tanks, motorcycles, army orchestras and 28 military planes sent to show of the superior military power of the Third Reich. There were only a few sporadic and poorly coordinated fire exchanges between German and Danish soldiers on Bredgade, among others, the street where the tram ran, and then the Danish state capitulated just a few hours into the morning. I have this memory image of my young grandmother bustling through Copenhagen inside one of the small tramcars in the old tram system above the ground, tears rolling down her cheeks as she told me. Passing through Kgs. Nytorv, where Hotel D'Angleterre had been turned into Nazi headquarters, and a flag with a swastika was raised side by side with the Danish flag. It is not my memory. Nevertheless, the image of my grandmother's memories is deeply intertwined with my memories.

Human power has a human memory. As said, many Europeans have memories from the Second World War passed down by parents or grandparents. Experiences and memory images of the type of human power that evolves through history into the persecution of one ethnic group emanating from class divides, economic struggles, and the decline and rise of nation-states. A violent, loud kind of power that kills and destroys. Like the shelling that killed family and friends in the living room of the mother of Jan Kleijssen, former Director of the Information Society at the Council of Europe, the heart of the human rights legal system in Europe.[19]

In 1944, during Operation Market Garden in the Nazi-occupied Netherlands, the houses in a British perimeter were shelled with German mortars and artillery. Jan Kleijssen told me that his then 17-year-old mother was in a living room full of people, neighbours, and friends in one of these houses. Ten seconds after she got up to get a glass of water in the other part of the house, a shell exploded in the room and killed everyone. Had the German gunner fired his gun ten seconds earlier, Jan Kleijssen would not have existed. He is one in that generation of Europeans marked by their parents' and grandparents' experiences of the atrocious kind of human power reinforced by advancements in science and technology. Germany developed the first jet engine during the war to power the Messerschmitt Me 262 fighter jet, one of the fastest planes of the time. They also developed the world's first long-range ballistic missile, the V-2 rocket, which was used to attack Allied cities in Europe. The first atomic bomb was created by the United States and dropped on the Japanese city of Hiroshima on 6th August 1945.

Fortunately, the human power to remember is also linked with the power to forget, forgive and move forward.[20] Thus, forgiveness and various forms of conflict resolution are core elements in religions worldwide and are incorporated into our legal systems. Forgetting and forgiveness are key to the functioning of a just and forward-looking human society. Only humans have this capacity, in stark contrast to the digital data archives of the 21st century. The Internet, for instance, was not designed to forget. But it must, and so we have had to redesign it; we had to create data protection laws to protect the private spaces of individuals, and courts in countries all over the world have exercised "the right to be forgotten" to grant individuals the right to have information about them removed online.[21] Still, today, AI systems thrive on not forgetting. They need the training data to evolve.

Jan Kleijssen's memories of war were a driving force in his career in the Council of Europe, aiming to prevent precisely the brutal type of power we carry with us in memory. He would "put heart and soul", as he expressed it, into the Istanbul Convention on preventing and combating violence against women and domestic violence, as well as a recommendation on the treatment of the children of imprisoned parents. At the end of his career in the Council of Europe, he led the development of one of the key global AI governance instruments, an international convention on human rights and the rule of law in the development of AI technologies.[22]

We tend to feel human power as a force held by the few and enforced with noise, violence, and oppression. Karl Marx described power as a limited resource held only by one social group at a time. It is upheld in a society governed by the needs and rules of the capitalist market. The ruling class holds all the power and uses it to exploit the working class. It's a type of power of the few and forcefully exercised in a factory, where workers work long hours for little pay and under harsh conditions to achieve a profit that reinforces the power of the economically ruling class.

On the other hand, the kind of power that the French philosopher Michel Foucault describes is more difficult to discern. Power is not external to us. It's not something you look at from a distance and can challenge from the outside. Not only expressed in sovereign state power in the form of the police and military. Power is diffused in everything we are as humans and in human societies, and it is held together by "regimes of truth", a form of internal power that "disciplines" us and coerces us from within.[23] This form of power ensures that we know what is true, how we should view the world and our relations with others, and how to behave and speak accordingly. Power is everywhere,[24] embedded in the way we communicate and construct knowledge, in education, in our treatment of the ill, in our sexuality, in the way in which we conduct science, and in public administration.

Human power has a human condition. We know that human power can be brutal. However, are we conflating human power with this knowledge of the forceful and oppressive use of force by the few? Focusing solely on the power of a select few could indeed lead us to overlook the vast potential of human power. Could it be that human power is boundless and unlimited?

The philosopher Hannah Arendt, who lived through the Second World War, says strength and domination are not power.[25] There is a human creative limitless power which is best expressed in undisturbed or "uncorrupted" human collective action. It is a power that all humans have, but it is at the same time also conditioned and constrained by our very human reality, which she describes in very few words:

> (...) life itself, natality and mortality, worldliness, plurality, and the earth.[26]

As simple as that. This is our human condition – the basic structure of human power. One can also imagine additional conditions where human power will flourish or perish, and we have already done so. War is one. A condition that, in a very obvious way, limits human freedom and well-being and thus empowerment. But how about the more invisible conditions of our everyday lives? Which types of social interaction do we prioritise, which are rewarded, and how do our social and material spaces condition these?

Take as an example education, the school as a space, and the role of the teacher and the students. In traditional education, teacher and student relations and forms of learning privilege learning through prescription, a classroom organised with school benches where students face the teacher and are placed on scales based on a system of numbers scoring the public performance of each student to ensure uniform behaviour. Alternative approaches to education, on the other hand, include different types of educational spaces, from sitting in "horseshoes" to moving around in open spaces with instructors to digital forms of asynchronous learning where students access classes in virtual spaces at different times. Also, learning and the educator-student relationship can be designed differently, putting a value on individual critical thinking and performance and rewarding students with qualitative feedback that does not score them quantitively in a uniform system in which they are compared with others.[27] Apply this imagination about educational spaces and interactions to our everyday spaces that may be designed to entice critical thinking and embrace a plurality of human voices and human traits such as creativity, compassion, and wisdom. Which conditions nurture human power?

Of course, it is not always this clear-cut. Creating specific human conditions that adapt to the complexity of human power in one way, or another is a difficult task.

In the late 1990s, open offices became the next big thing in office planning. Companies and organisations worldwide started pulling down walls and redesigning office spaces to create open spaces to enable cultures of open collaboration among their employees. However, studies soon showed that what had happened was, in fact, the opposite. Productivity was lower in open workspaces.[28] The humans were adapting to their new conditions, but not as expected. They created "social" walls around themselves by, for example, wearing headphones and policing the open space when creating new social rules based on reading the signs from each other on when to be left alone. Creating an open office environment often results in less direct, meaningful interaction between colleagues than more. One study conducted at two Fortune 500 companies before and after the companies transitioned from cubicles to open offices, for example, showed a 70% decrease in direct face-to-face interaction.[29] This was not the desired open, flexible and productive outcome the original designers of open office spaces had imagined.

Today, we tend to privilege and prioritise more interaction, thinking and creating together, voluminous and extrovert performance in a "world that can't stop talking", as Susan Cain describes it in her book *Quiet*.[30] This "New Group Think movement", she says, is embraced by corporations, in particular, but also more generally, we increasingly have to exist in spaces that are not "cultivating solitude".[31]

Susan Cain believes introverts have a particular power that these environments do not respect and respond to. Nevertheless, according to Cain, social media has made life easier for introverts who have found a way to express themselves without direct interaction with other humans. But could we not also think of social media differently? Perhaps social media represents precisely the kind of space that works against solitude, providing eternal access to the "world that can't stop talking".

The "Slow Web Movement"[32] offers an alternative to the "Fast Web" that we are most accustomed to and which the science fiction writer Jack Cheng describes as "frenetic" and "addictive", compact with an overload of communication in "real-time". The slow web, instead, he says, is more attuned to human slowness and sensibility, with the information presented to us in a "timely" manner when we have time and space to absorb it. The "randomness" of the information we receive in one compressed and extended space with continuous access through our mobiles, computers, and connected devices, in general, is with the Slow Web replaced with interactions based on our individual daily human "rhythms".

In *Data Ethics of Power*, I explored the big data and AI socio-technical infrastructures of power in our everyday lives and politics and concluded that: "(…) human power needs specific spatial and temporal conditions to flourish (…)" why we urgently need to:

> (…) actively build alternative sociotechnical data infrastructures and systems that interact with human agency, power and ethics in a different way (…).[33]

If we return to Hannah Arendt, she considered contemplation and independent critical thinking core preconditions for the kinds of human resistance and political action that can challenge totalitarian regimes of power such as that of Nazi Germany. She believed these human capacities and the capacity to do evil were not inherent to individual human beings. They are conditioned. For example, she thought solitary withdrawal from public space rules and common voices was necessary to nurture contemplation and independent critical thinking. Arendt was deeply impacted by her experience of attending the SS officer Otto Adolf Eichmann's trial in Israel in 1961, which led to his judgement and execution in 1962.[34] She saw that he was not an evil mastermind acting out of his evil nature. He was pretty ordinary, even unintelligent, she said. Eichmann's evil deeds resulted from an inability to analyse and assess the specific circumstances of his time rather than any inherent malevolence.[35] His "evilness" was banal because he acted within the conditions of the prescribed Nazi worldview and ideology. In fact, she argued, he was not a monster, alien to human nature, but a human being without critical faculties. As the fictional Nazi professor, Siletsky tells the actress Maria Tura in the 1942 movie *To Be or Not to Be* when trying to change her mind about becoming a Nazi spy:

> We're not brutal, we're not monsters (…) Tell me, do I look like a monster? (…) We're just like other people. We love to sing, we love to dance, we admire beautiful women. We're human. And sometimes very human.

The real Nazi officer Otto Adolf Eichmann was also one kind of human being among other human beings in an eroded and distorted public space and order.[36]

Let's think about this: Could it be that digital transformations, AI, and big data are eroding the qualities of public space; the conditions for human power today? We can think of our everyday lives as spaces conditioned by not just doors, walls, streets, and bridges but also virtually, with digits and cables. We are constantly interacting with digital spaces, data, and analytical tools and just like the doors we pass through every day, they may provide us access to things or, if locked, limit our access. Like streets, these digital spaces and their infrastructures will lead us in specific directions. The virtual spaces humans navigate are like commonplaces that blend into the background of our daily lives and go largely unnoticed yet play a vital role in organising and facilitating our routines. We find ourselves in these everyday life conditions that can give us strength and inspiration but also limit human power.

Philosopher Lenka Ucnik draws lines to Arendt's descriptions of the conditions for human political action in the social media age. In Arendt's understanding, Ucnik says, the design and delivery of social media platforms, such as Meta, former Facebook, have contributed to the destruction of the necessary preconditions for genuine deliberation.[37] Tailored newsfeeds, for example, aim to increase user engagement and profit margins but create different "realities" and "truths" for different users, eroding the plurality of views. Legal scholar Nathalie Smuha also draws on Arendt to illustrate how our ability to lead a fulfilling life through meaningful connections is significantly impacted by the widespread application of AI systems on, among others, social media platforms today.[38] Another legal scholar, Julie E. Cohen, argues that our online conditions diminish the privacy of individuals and consequently limit humans' capacity to be creative and innovate. Innovation, she says:

> ...requires room to tinker, and therefore thrives most fully in an environment that values and preserves spaces for tinkering. A society that permits the unchecked ascendancy of surveillance infrastructures, which dampen and modulate behavioral variability, cannot hope to maintain a vibrant tradition of cultural and technical innovation.[39]

Conditions prioritising privacy, freedom of expression and human connection are prerequisites for individuals' ability to act and think independently. Many studies show how surveillance and the feeling of being watched online stifle human power – our ability to innovate, think critically and connect with other human beings.[40] Thus, as Tranberg and I have argued, "Privacy is empowerment"[41] and hence conditions where we are "not left alone" are at the same time hampering fundamental human powers.

When I spoke with Denmark's Tech Ambassador, Anne Marie Engtoft Meldgaaard, in 2023, she at some point reflected on 1970s science fiction literature visions of what technology can do to liberate humans and bring us closer. These were, of course, utopian ideas, she said; nevertheless, the depiction of how technology can shape the conditions of humanity for the better would be more constructive than what we have now, she said – also for technology politics:

> It is better than the feeling that we have about technology right now, that technology is taking us further away from each other. If we have this dialogue in a space where it isn't just about how many phones I'm going to sell, how many minutes I'm going to make you spend on my social media platform, or how we can win the geo-political technology race. It is about what kind of power we've created with technology instead, about human needs and how it can benefit as many people as possible; I think of this space almost as a wonderful summer camp. A beautiful place in nature where we can withdraw and try to work through our fundamental understanding of the role of technology without being disturbed. I, for example, see spaces like that created around feminist technology. They are almost like the contemporary sci-fi room, where researchers make room for dreaming about what technologies can do for women, gender equality, and what it will mean for the world. I can imagine rooms like these created for green technology solutions, health tech, and many other areas. It may seem naive, but I believe in the best of humanity and that technology can be a medium for that, but you need a free space to accomplish that.[42]

Human power is a force of nature. In the dance performance *Act of Gravity*,[43] the dancers' bodies and movements are conditioned by Earth's gravity. In the endless darkness of space, they graciously, with some visible effort, bounce off the stretched-out net of a trampoline, touching the wall that seems to be rising out of nothing and are then pulled back down. Arms circling, legs running in the air in search of the surface. "Who are we humans eternally falling while the laws of physics set everything into motion?" choreographer and artistic director Tina Tarpgaard asked when I spoke with her in 2023.[44] Human power is like gravity. It is within us and around us, indistinguishable from very material surroundings, and we learn to dance with it graciously but not without effort. We bounce in and with it, struggle with it, embrace it, are pulled by it, and retract from it. Gravity is a form of external power, the condition of the human dancer, but it is also the internal force of the dancer. Her creative human power forcefully navigates the power of gravity with movement. As Tarpgaard said:

> As dancers, we negotiate gravity all the time; it affects us all the time. That creates the dance. So, it's so fundamental. It's the cornerstone of being a dancer.

When her daughter took her first great gasp of air after the delivery from her womb into the hands of people and the open space of a hospital room, Tarpgaard was inspired to create a dance performance based on breathing. The dancers' bodies turn and twist with and around their breathing. In another performance, she worked with software programmers who combined motion-tracking technology with light installations. One dancer creates a large surface of light, like a piece of paper, with his body. The human feeds thousands of mealworms with flamingo cardboard in her latest art installation. She said that when they digest the plastic material, it sounds like when oil burns on a frying pan. Here, she wanted to put together two very different agents. The human and the mealworm. An animal that most of us are disgusted by. One that feeds on rotten and dead things. But even that creature we can interact with in a frictionless way.

Human power is not unbounded but subject to conditions and always in interplay with other forces. Breathing, gravity, light, and other animals. We may be in harmony in our interaction with these conditions and forces, as Tina Tarpgaard wants to show us with her art, or we may be in conflict bodily, as when a virus invades the body during a pandemic, or mentally, when the quiet and contemplative psyche suffers in a world that "can't stop talking".[45] However, as Hannah Arendt writes, these conditions" cannot explain who we are because they do not condition us completely".[46] Each human being is uniquely placed in time and space. Very diverse human experiences shape our human conditions. No dancer will experience space and navigate with the pull of gravity in the same way Tarpgaard explained. All the dancers, for example, when navigating space, will use their unique experiences about gravity as acrobats, ballet dancers, or contemporary dancers.

Human power is like music. Harmonic or disharmonic powers are generated by the amalgamation of humans with situated and collective experiences and their conditions. Arendt talks about the power of the "in-between space" between humans. This is a space where we act as individuals uniquely placed in our own time and space, sustained by personal memories that may be transposed through generations but also connected through our standard conditions and memories shared with other human beings. As Leo J. Penta writes about Arendt's idea of power:

> Power, unlike strength, force, or violence (from which Arendt specifically and unequivocally differentiates power), is neither a property nor an instrument, nor any sort of monadic phenomenon, but rather has its existence inter homines when people act together in public.[47]

When Jan Kleijssen told me about his mother during the Second World War, I got a flash image of my young grandmother on the tram in Copenhagen. These are two different experiences that were somehow transposed into an "in-between space" that generated our shared sentiment – an aversion against totalitarian forms of power. At the same time, these simultaneously shared, distinct human memories and contexts of political action constitute forms of power that we will never be able to represent and transform into direct action as they can only exist in a distinct human context.

The "in-betweenness", or the movement, of a time that is not experienced as a "before", "now", and "after" is a time that, like Henri Bergson says, can only be experienced as it happens, never be re-lived or represented (see also Chapters 1 and 4). We will never be able to fully represent the memories of our parents or grandparents of the Second World War with images or words. Nonetheless, memories like these can be human drives for action. This form of lived human power might also become drivers for a type of collective action, what Bergson refers to as an "open love" (see also Chapter 5), where differences are not posed against each other as conflicting opposites but combined as in a piece of music:

> Imagine a piece of music which expresses love. It is not love for any particular person. Another piece of music will express another love. Here, we have two distinct emotional atmospheres, two different fragrances, and in both cases, the quality of love will depend upon its essence and not upon its object. Nevertheless, it is hard to conceive a love which is, so to speak, at work, and yet applies to nothing.[48]

Let's imagine human power as a piece of music that is most beautiful when each musician finds harmony in their distinct individual and, at the same time, collective human experience. Musicians in an orchestra are highly interdependent. Intonation in a music group depends on the musicians' ability to create harmony by balancing each other's unique qualities. The musician Uffe Savery, best known as the drummer in the 1990s hugely famous musical duo Safriduo, describes this musical process as "deep listening and interdependence". You don't *move* in on each other in music; "You *listen* in on each other", he says.[49]

Could we imagine human power in politics in the same way? In the 70 years of the Council of Europe's history, every single achievement, from the European Convention on Human Rights to the AI Treaty, all the social changes, Jan Kleijssen says, have always happened because of and in between people:

> Staff members of the Council of Europe turned out to be driving forces, building allies with politicians, with certain governments, and also ministers who got involved. In all of this, it's always been people who drove change.

We can imagine human power as a collective human drive fuelled by individual lived memories that work together in "in-between spaces", like an orchestra, to achieve change. This is the power of humans that we can design machines to reinforce, but they can never replace. A power like that needs specific conditions to thrive. Let's consider the conditions of human power in our most recent history. How they evolved, and which social trends, human hopes, and dreams were invested in them. We may explore two competing social trends in more depth here: one that is fuelled by an ardent faith in the unlimited potential of technological power, another by fiery convictions about the potential of humanity, human power and freedom.

BUILDING THE SOCIETY OF THE DESTINY MACHINE

The World Exhibition was held for the first time in 1851 at the Crystal Palace in London, with the Industrial Revolution on display. Hydraulic presses, printing machines, and steam engines were just a few of the inventions exhibited at the first World Exhibition. Since then, World Exhibitions have been taking place worldwide regularly, with human excellence on display conveyed in the shape of the Eiffel Tower, ice cream cones, Ferris wheels, telephones, x-ray machines, televisions, touch screens, and many other extraordinary inventions. Unique exhibitions taking place every five years in countries all over the world. However, one thing more powerful than the human thrives for excellence is a pandemic, which is why the World Exhibition, which was supposed to take place in 2020 in Dubai, the first year of the COVID-19 pandemic, was instead held in 2021–2022.

Dubai, a city in the United Arab Emirates, rose out of a desert. At first, it was a small 18th-century fishing village, but it grew into a metropolis. With the discovery of oil in the 1960s and extensive trade, it rapidly turned into a city of wealth, skyscrapers, beaches, business, and tourism. As such, it is the perfect place to display what humans excel at when mastering their environment and building extraordinary things out of practically "nothing". One should be impressed.

Still, visiting the World Exhibition in Dubai in March 2022 left me with a feeling of uneasiness. Memories of the event and the place in which it took place are to me today intertwined with the awkward smile of the artist that was brought to Dubai from Rome sitting across from me at a Chinese restaurant looking at the world's largest musical fountains, the Dubai Fountains, vivaciously swaying and bopping to the sound of a range of deafening musical ballades. The unidirectional streams of humans crossing the bridge by the fountains restrained by water concrete and lid-up buildings. It actually felt like the human striving for excellence was outperforming its own project's very human identity.

At Expo Dubai, human crisis and culture were compressed and immobilised in huge, obscure buildings. The Spanish pavilion, the Finnish pavilion, the Italian pavilion. I quickly snapped a picture of the white Serbian pavilion for a friend while passing by it in a buggy. Driving by the Russian pavilion, a huge organically formed building of coloured strings of wire, a person next to me whispered, "We need to take a picture; this will probably not be here next time". So, simply put. So swift. Russia had just invaded Ukraine. My mind travelled out of place, and a mental image of tanks, the ruins of bombed buildings, human faces twisted in pain, and lifeless bodies were transposed onto the pavilion, and then it was gone. We had already passed it. At Expo Dubai, we only saw progress and excellence. The latest inventions in aerodynamics and aviation, AI curation, holograms, virtual reality, and talking robots. No human struggle, destruction, or mess that could not be remedied with an excellent piece of technology. Human ambivalence appeared only in the cracks: in the sweat of the solid faces of the streams of people at the hot site of Expo Dubai, a 438-hectare area located in the desert between the cities of Dubai and Abu Dhabi, or in the voice of the Indian American activist exhausted and with painful feet after getting lost in the labyrinth of Expo pavilions and paths. She told me about the Indian people bicycling for life, not recreation, kilometres and kilometres, from city to city to get medical help during the pandemic.

Expo Dubai was one big human mess distilled and immobilised in plastic, presentation, and politics – a Disney land for adults. I tried to remember – how was it now that Jean Baudrillard put it?

> Disneyland is there to conceal the fact that it is the "real" country (…) It is no longer a question of a false representation of reality (ideology), but of concealing the fact that the real is no longer real, and thus of saving the reality principle.[50]

This World Fair, created to exhibit human excellence, conveyed more than anything the ardent faith of – some 21st-century scientists, many policymakers, and most of the global industry – in the unlimited potential of technological power. It was also not only a "representation", a display of dreams and hopes about the destiny of humanity, this World Exposition was the real thing; an extension of a society expressed in technologically dominated human interaction. In this society, technology is not just a tool. It has become the embodiment of human power, what we believe humanity excels at and is our most unique capacity. Here, we increasingly perceive humanity and human power through technological power. Thus, we do not develop technologies to mirror human power, we design them in the image of their own technological imitation of human power. A car is no longer a means of human transportation, it's the adventure, and the safety that the technology promises. A phone is no longer a human means of communication, it's the world and the network. A camera is no longer a means of human recollection, it is the memory and the perspective.

This is, what I elsewhere have called "The Society of Destiny Machines and Algorithmic Gods",[51] which is characterised by a complex and advanced machinery that leads, guides, and defines our lives. The "Destiny Machines", such as social media platforms, online search engines, generative AI systems for images, text, and video, predictive policing tools, and public tax or welfare benefit distribution systems, are not just new technologies. They are businesses, organisations, and scientific labs occupied with the destinies of human beings. The products and services they develop are thus mainly designed to predict human behaviour and act on these predictions using the massive amounts of data made available by digital transformations.

What is more, these Destiny Machines are integral to our 21st-century socio-technical infrastructures of power. We may here consider two types of "infrastructures of power" of the digital transformation: "Big data socio-technical infrastructures of power" and "AI socio-technical infra-structures of power".[52] Big data socio-technical infrastructures are fuelled by big data technologies. They sustain global economies and societies across different geographic territories, legal jurisdictions, and cultures and have evolved into the background against which all our social practices, identity construction, and politics are conducted. "AI socio-technical infrastructures" are an evolution of the big data infrastructures with AI components that are often designed to sense the environment in real time and learn and evolve with autonomous or semi-autonomous agencies. While the big data infrastructures of the spaces we move around in have transformed all data (including the data retrieved from and on humans) into digital data, the AI-infused infrastructures act on that data to actively shape our past and present in the image of the future. This also implies the moulding of human life into immobile entities that can be worked on, analysed, and predicted.

It is these conditions that are enabling the Destiny Machines and their services that produce, create, and define the destiny of human power by interlocking every individual's human life with prediction. Every day, whether we like it or not, algorithms are deployed to predict our behaviour based on the data derived from what we do now and what we have done before. Our lives are framed, and we are pointed in specific directions. That is what Destiny Machines do. They produce machine-readable people, represent human power in an interoperable form, and spit out destinies on the other side of the production line. And they evolve in invisible contexts of powerful interests that ask us to accept our produced destinies without questioning the interests behind them.

Human lives are a resource of the Destiny Machines' machinery. A company like WorldCoin may speak passionately about a fairly distrib-uted, cryptocurrency-based universal basic income while at the same time sending out scavengers to 24 countries, 14 of those developing nations, to scan 450,000 eyes, faces, and bodies of people who are paid nothing or very little to train WorldCoin's neural networks. It is a machinery that is reinforced in its complexity by exploiting human power, as one of the investigating MIT tech reporters, who revealed the WorldCoin practices, says:

> The massive effort to teach WorldCoin's AI to recognise who or what was human was, ironically, dehumanizing to those involved.[53]

Paradoxically, the core challenge to human power is that Destiny Machines cannot handle its complexity. They are designed only to identify and act on the symptoms of human lives. The flagship health AI of IBM Watson Health failed precisely for this reason. It was sold in 2022 to venture capitalists for a price, although undisclosed, that was speculated to be a quarter of what it had cost IBM in acquisitions since its launch in 2015. Watson Health would analyse patient symptoms to find the most probable diagnosis. Still, it would, time after time, score worse than human medical practitioners because it did not see the whole patient history and the context of care.[54]

Evidently, individual human beings today have little awareness of the conditions of the Society of the Destiny Machines and increasingly have even less freedom and power to decide the direction in which it takes them. This doesn't mean that we cannot live, that human life does not continue, and that human society does not evolve. But it does mean that we are less in charge and have less freedom, and it might also mean that the character of our human power is transforming. It is like the universe of the "holistic detective" Dirk Gently portrayed in Douglas Adam's 1988 fantasy novel, *The Long Dark Tea-Time of the Soul*. Although inexplicable to human comprehension, Dirk still manages to solve mysteries across peculiar dimensions and eras, unhesitatingly acknowledging his place in the universe's odd workings, trusting that it will consistently position him precisely where he must be. As he says:

> I may not have gone where I intended to go, but I think I have ended up where I needed to be.

This is a gloomy depiction, but the very human feeling of disempowerment that the Society of the Destiny Machine triggers in all of us is also a human power. According to a YouGov Poll from 2024 on how Americans feel about computers and human intelligence, close to half of all respondents fear that AI could attack humanity and the majority believe that human intelligence will be overtaken sometime between 2025 to 2099.[55] Fears about the mechanical replacement of human powers challenge us to revisit our most set ideas about what it means to be human. Are we just noise in a perfect world? Will AI outcompete our less-than-perfect human creativity, intelligence, and skills and replace us in the workplace, in art and science? These are feelings; however, as you will see in Chapter 2 on emotion and in the last part of this book on emotional politics, human feelings are integral to human power. Thus, we should therefore also consider the existential fears that, in particular, generative AI, have triggered, a form of power. After all, this gut reaction of fear is still very different from Dirk Gently's indifference, and we could therefore, based on our very human fears, try to find the answers to questions about how we got here.

Certainly, reflection and promotion of human power have not been absent in the history of humankind. In the recent 20th-century history, ideas about the conditions and restraints of human power have even been the foundation of some of the most critical social and cultural movements.

HIPPIES AND POSTMODERNISTS

In parallel with the early evolution of the Society of the Destiny Machine, we have in the 20th century seen the emergence of social and cultural movements based on articulate ideas about human power. Fuelled by a sense of human disempowerment, and despite their varying expressions and occasionally conflicting messages, a common thread ran through these counter-culture movements: humanity is inherently complex, and its potential and power can be realised above all through human freedom.

Not long after the Second World War was declared over, the French philosopher Jean-Paul Sartre caught attention with his public address, "Existentialism is a Humanism". He described the loneliness of the "atheist" when realising there is no God to rely on for life guidance and prescription, no external power to provide moral certainty or life direction. Every human being is responsible for their existence, he said. However, not only for themselves but for humanity.[56] The German Philosopher Friedrich Nietzsche had proclaimed God dead in the previous century, in fact, with some concern as he believed that the enlightenment had left humanity scattered with no firm belief system or worldview.

Yet, in the years following the Second World War, many secular bodies were set up in the Western World echoing Sartre's humanistic outlook. With roots in the humanistic tradition of the 19th century, a "free-religious, free-thought, ethical, atheist and rationalist"[57] worldview was adapted to 20th-century social and cultural changes to advocate human rights and democratic values and their institutions such as the United Nations.[58]

Other cultural and social trends expressing variations of a humanistic narrative have been influential during the 20th century when confronting dominant reductive social prescription that essentially reduces human complexity and, consequently, human beings' inherent power – the prescription of adults, society and the state, knowledge, and its grand narratives. Let's look at two movements like that: the youth countercultures of the 1960s–1970s and the postmodernist art, film, and theory of the 1970s–1990s.

The 1960s–1970s "counter-culture", also called the "hippie-movement", started in the United States as a youth rebellion against an adult society's moral and political prescription. It was expressed, among other things, in anti-Vietnam war protests with the infamous message "Make love, not war" and spread across different, primarily Western countries, promoting human freedom when challenging established conservative social norms of the adult generation with everything from alternative clothing and forms of living to an exploration of the self with psychedelic drugs.

My uncle Bo Erik Hasselbalch lived through the period and carried with him its core humanistic identity articulated in a spiritual and holistic approach to life, which profoundly influenced everything from his choices in healthcare to his business relations running a successful film school in Copenhagen. When I asked him about his experience at the time, he said:

> In hindsight, the period was not conducive to much other than the unleashing of creativity, such as music. We saw what was wrong but had no idea what could replace the old. The period was very idealistic, while most of us rushed out and made mistakes. Everyone today has strong opinions about the time. For me, the era revolutionised the rest of my trajectory. But it might be explosive for you to see the youth rebellion in any clear-cut way.

Then he told me to watch the 1970 movie *The Zabriskie Point*.

The Italian filmmaker Michelangelo Antonioni visited the United States in 1967–1968, and based on, among others, his experiences of the demonstrations at the Chicago Democratic Party convention,[59] he decided to make the film. The movie's plot was based on a newspaper article he read about a young man who stole a small aeroplane in Arizona and was killed by police while returning it. Watching it today, I see an unsentimental depiction of the counter-youth culture of the time. This is probably also the reason for the movie's bad public reception in 1970 when it was first released, described by some critics as Antonioni's worst movie ever.[60] I see the movie's qualities differently. Apart from its point of departure in a youth movement, it does not seek to tackle topics such as the Vietnam War directly or even race and gender equality. However, it does bring about a very accurate picture of human struggle – not perfect, not coordinated, and perhaps not even very constructive, of humans striving to break free from established settings where they no longer feel they belong. It provides a picture of the contrasts of the society the young "hippies" lived in and with. From war (violence) to love. From culture and artifice in school, work, family life, and the cities to the authenticity of humans in nature. In reality, the movie's narrative is insignificant, and if this is where you seek an understanding of the period, you might agree with its critics that it provides very little content on that. That said, the contrasts of the time are found in the scenes of the movie that pan from a university where the main character, Mark, leaves from a meeting where students are planning a student strike because he is "willing to *die*, but not of *boredom*" to a clash between students and police resulting in violent mass arrests. It continues into a police station. Then, it moves effortlessly into an office building where a real-estate executive in his perfect luxury business surroundings prepares for a real-estate development in the desert – panning back to the police-student conflict where students are teargassed and shot, and one police officer is killed. Away into the air in the aeroplane that Mark steals. Then, back on the ground, away from the violence, the rigid contours of the city and office buildings to the organic forms of the brown desert that the second main character, the girl Daria, who is working for the real-estate executive, is driving through. The two young people are united in the desert, where they make love (not war) in the sand among other naked, unidentified young people. Then, away again, in the aeroplane, which is now painted with political slogans of the youth movement. We return to the real-estate executive's excessive surroundings when Daria drives to his luxury house after learning that Mark has died. She stays a little in the extravagant house inhabited by shallow women by the swimming pool, and the executive is busy in a business meeting. When finally leaving the house, she turns around to look at it one final time. In her mind, the house and its items of a consumerist lifestyle explode in slow motion.

Human crisis is very real when conditions restrain us. Still, with this explosion, we are also presented with the human potential to break free, reminding me of one of my favourite quotes by philosopher Henri Bergson that I will refer to many times in this book:

> In vain we force the living into this or that one of our moulds. All the moulds crack. They are too narrow, above all too rigid, for what we try to put into them.[61]

Core ideas about the untapped and restrained human potential ran through the various strands of the countercultures of the 1960s and 1970s. One was the "human potential" movement carried forward by figures such as the psychologist Abraham Maslow, who saw a hidden potential in each person just waiting to be unlocked.[62] The human potential movement constituted a turn to humanist psychology and existential philosophy. It was expressed in everything from practising Eastern spiritual traditions, experimental theatre, and encounter groups to alternative organisational development and training of soldiers and spies.[63] As a counter-reaction to what was perceived as the artifice of previous generations and traditions, the human potential movement comprised a search for a type of human "authenticity" in what was perceived as a humanly disaffecting society. Searching for the "authentic" also meant exploring human nature, the inherent potential of humans, the "original" as opposed to the derivative, which could be expressed only in "true" human emotion and feeling. As the son of J. L. Moreno, one of the human potential movement's key figures, Professor Jonathan Moreno writes in the book about his father:

> Authenticity meant being honest and open about one's desire for love and acceptance in an increasingly alienating, corporatised, technological, violent, and materialistic world. And it meant being willing and able to give love in return.[64]

The counter cultures of the period take form as a flight from the ground of everything that binds us, like the seagull Jonathan Livingstone in Richard Bach's 1970 novel that cannot stop flying as freedom is its "true" nature:

> It is right for a gull to fly, that freedom is the very nature of his being, that whatever stands against that freedom must be set aside, be it ritual or superstition or limitation in any form.[65]

The postmodernist cultural movement in art, filmmaking, and critical theory of the 1960s–1990s is another "countermovement"; although, the search for authenticity and true human nature is here an oxymoron. There is no true human nature or state of being that we can reach beyond what we already have. Unlike the political ideology of the 1960s and 1970s youth countermovements, there is no "true" grand narrative to guide humanity. Nevertheless, the postmodernist shares with other humanistic strands of the same period the struggle against inherent power dynamics of previous generations' social practices and institutions. The nuclear family, the state, the factory, the museum, the school, and the ideology – all represent the prescription of what is perceived as an external humanly restraining order. After all, Modernity, as the historian Paul N. Edwards describes it, was embodied in a human experience and a "lived reality" of control and order:

> To live within the multiple, interlocking infrastructures of modern societies is to know one's place in gigantic systems that both enable and restrain us.[66]

Andy Warhol's "Campbell's Soup Cans" (1962), consisting of 32 soup cans, tackles the industrial age's mechanical mass production and consumer culture. Meanwhile, Roy Lichtenstein's series of comic book-inspired paintings "Whaam!" (1963) challenges elitist notions of the original artist and the social prescription as to what constitutes "real art". Postmodernism is also a realisation that there is no "outside". There is no moral guidance to rely on outside our human representation systems – language, images, and social structures. We are as integral to these as they are to us. David Lynch's movies' "intertextuality" creates a universe internal to the context of film production in which they exist, with references between his movies and other movies. They draw their characters, actors, messaging, and narratives within this internal universe of films and the film industry. In his 1990 movie *Wild at Heart*, the Hollywood blast from 1939 *The Wizard of Oz*, produced by the mega film production corporation Metro-Goldwyn-Mayer, is the main frame of reference. The couple, Sailor and Lula, have a similar narrative of self-discovery as that of the *Wizard of Oz*'s Dorothy when they try to escape Lula's mother, the Wicked Witch of Oz, and the violent criminals pursuing them. The Good Witch appears in the embodiment of Lynch's Twin Peak series main character, Laura Palmer (played by the same actress), in Sailor's pink vision to guide him with the words: "If you are truly *wild at heart*, you'll fight for your dreams. *Don't turn away* from *love*" paraphrasing the words of the Wizard of Oz to Dorothy: "And remember, my sentimental friend, that a heart is not judged by how much you love; but by how much you are loved by others".

When insisting that there is no "outside" to be guided by or rely on, the postmodernist also recognises that the sense of human disempowerment that Modernity represents must be challenged from within. We only have ourselves and what we have created. No outside force or principle can solve our human crisis. This postmodernist recognition of what we could also call a human fundamental "loneliness" is a form of human power like that of other more obviously humanistic movements of the 20th century. All we can rely on is human power and its condition, which is not only a frightening realisation but also a form of human liberation.

Famously, the French philosopher Jean-Francois Lyotard, in his famous book *The Postmodern Condition*, published in 1979, challenged blind trust in scientific knowledge as univocally true. He said that our knowledge, the way we know things and what we know, is just a type of discourse, which also means that it can change when its conditions transform. Lyotard described a "computerised" society where human knowledge can only survive by fitting "(…) into the new channels, and become operational, only if learning is translated into quantities of information".[67] We can thus also expect, he further stated, the "exteriorisation of knowledge", which finally amounts to a loss of power – of the self, of the nation-state. This is not necessarily bad as the challenge to knowledge is also a challenge to what Lyotard calls a "paradigm of progress in science and technology, to which economic growth and the expansion of sociopolitical power seem to be natural complements". This basically means that old forms of dominant powers, such as "nation-states, parties, professions, institutions, and historical traditions", lose their attraction. Their "grand narratives" and the enforcement of these are contested.

The question is, when all that we know is just the discourse of dominant power, as the postmodernist claims, then where does that leave human power? Are humans powerless? Lyotard says no. Even the least privileged human being can counter dominant power by refuting the grand narratives when there are conditions in place for "speaking" differently and countering with alternative language:

> Does the university have a place for language experiments (poetics)? Can you tell stories in a cabinet meeting? Advocate a cause in the barracks? The answers are clear: yes, if the university opens creative workshops; yes, if the cabinet works with prospective scenarios; yes, if the limits of the old institution are displaced. Reciprocally, it can be said that the boundaries only stabilise when they cease to be stakes in the game.[68]

Ardent visions about human power, the human inherent potential, human freedom, and responsibility have thrust ahead some of the most influential 20th-century social and cultural movements. Contrary to what the Society of the Destiny Machines dictates, human fate and the destinies of human beings are, in these views, not predetermined. Humanity is complex and dynamic, and our future is open. Nevertheless, in the early 21st century, it seems that these humanistic views on an open human future have not prevailed. How could they? In a society where a new type of dogmatic agencies, the "Algorithmic Gods", are every day working in very concrete ways on the destiny of humanity, interlocking it in past data with predictions, and prescribing where each human being is going.

ALGORITHMIC GODS

Over two millennia ago, Marcus Tullius Cicero, a Roman philosopher and statesman, presented the term "Humanitas" (from Latin: humanity, nature, and human dignity). This term later became foundational to 15th–19th-century humanism, where it was used to describe the inherent potential and power of human beings. Humanitas represents the virtues and qualities human beings cultivate when exercising power with empathy and compassion towards others. These virtues and qualities are not formal doctrines; instead, they are internalised by human beings, granting them the ability to govern their moral existence.[69]

On the other hand, Cicero also said that "physics" should determine the fate of human life, not "superstition":

> By "fate", I mean what the Greeks call heimarmenê – an ordering and sequence of causes, since it is the connexion of cause to cause which out of itself produces anything. (…) Consequently nothing has happened which was not going to be, and likewise nothing is going to be of which nature does not contain causes working to bring that very thing about. This makes it intelligible that fate should be, not the "fate" of superstition, but that of physics, an everlasting cause of things – why past things happened, why present things are now happening, and why future things will be.[70]

Since early human history, different conceptions of fate and destiny have challenged human power – from metaphysical justifications based on the reasoning of gods to those grounded in the laws of physics and the promise of progress and objective science. One faith replaces another, and while Cicero rejected superstition as our human destiny, he still accepted the idea of a human fate based on the laws of nature, thus binding human life to a predetermined direction and future.[71]

Now, let's consider this: Has "technology" replaced the laws of physics as the dominant faith of the 21st century? Scientific discoveries and developments have been testing and challenging our understanding of the laws of physics at an increasing rate over the last century. The discovery of dark matter and energy challenged our fundamental understanding of gravity, while quantum mechanics questioned unified theories about the nature of spacetime. The exploration of Mars, including missions such as curiosity and perseverance, has raised the question of whether we are alone in the universe as we search for signs of past life on the Red Planet. When the laws of nature are no longer a given, are we now instead leaving our fate at the mercy of the Algorithmic Gods invented by primarily the tech gurus of the 21st century?

At the beginning of 2023, in an article written for the World Economic Forum's annual meeting, Patrick Paul Gelsinger, the CEO of one of the world's biggest multinational corporations, the technology company Intel, described five "tech superpowers": computing, connectivity, infrastructure, AI, and sensing. These powers, he wrote, would transform human progress:

> We are just beginning to tap the full potential of 5 extraordinary tech superpowers.[72]

It is not the first time that an industry trailblazer has described various types of "technology" as the unswerving extension of human power. Every year, various versions of emerging technologies are presented as the new human superpowers that will univocally ensure our progress, unleashing human potential once and for all. This also means that we increasingly imagine the shape and direction of our human future through technology and trust in its univocal potential to empower and guide humankind.

I want to argue that technology has become the contemporary manifestation of human destiny, similar to the social dogmas that the 20th-century humanistic social and cultural movements were challenging. We are once again, as we have done throughout history, relinquishing our ability to shape the human future by placing our faith in external forces. By doing so, we are limiting the human potential and freedom, which also means avoiding the responsibility of determining the destiny of humanity. Is it not time for a renewed mobilisation of the humanistic movement? One that does not just promote and extend the interests of the tech gurus who created the problem, but one that takes off with people and civil society organisations and with a genuine interest in the uniqueness and potential of human power.

No doubt technology extends human powers. We build things and develop technologies to be useful and meaningful in our human lives, and often, they become essential to economic, social, cultural, and individual human powers. Glasses enable the weak-sighted to see; medicine makes us healthy and stronger; trains and aeroplanes take us across the globe, and the Internet connects the ones with online access to others with online access. Throughout history, the evolution of human science and technology has transformed and expanded the opportunities and powers of human beings. However, not in a completely frictionless way. As the media scholar Marshall McLuhan said in 1964, the technological "extensions of man" also leave us with complex questions that cannot be answered in any straightforward manner. In particular, the final phase of technological development that he thought we were moving into at that time, where we finally extend our consciousness, necessitates such human querying:

> Rapidly, we approach the final phase of the extensions of man – the technological simulation of consciousness, when the creative process of knowing will be collectively and corporately extended to the whole of human society, much as we have already extended our senses and our nerves by the various media. Whether the extension of consciousness, so long sought by advertisers for specific products, will be "a good thing" is a question that admits of a wide solution. There is little possibility of answering such questions about the extensions of man without considering all of them together. Any extension, whether of skin, hand, or foot, affects the whole psychic and social complex.[73]

"SIMPLE MACHINES" AND "SUPERTOOLS"

Here, well into the 21st century, amid an AI machine age, it seems that we are experiencing what Marshall McLuhan predicted in 1964 – the advancement into the final phase of the "extension of man".[74] Though, without the comprehensive approach that he also emphasised, we need to have. In fact, the direct technological "extension of man" has been integral to the history of the biggest technological hype of the 21st century – AI: Artificial

Intelligence. From the beginning, AI was designed to replicate a limited perspective of human capabilities, treating human intelligence, and cognitive abilities as traits that machines can imitate or even directly surpass.

The term "Artificial Intelligence" was coined in 1956 by mathematics professor John McCarthy, who organised the founding event of the AI research field, the Dartmouth Summer Research Project on Artificial Intelligence.[75] At this meeting, a group of pioneer mathematicians and scientists gathered to explore the computational imitation of the human brain (i.e., computer processes similar to humans when thinking with and learning from information) and to explore the automation of computers.[76]

Only a few decades after the Dartmouth event, AI systems with some capacities for autonomous decision-making were already being developed and deployed in various real-world settings. The 1980s "expert systems" were, for instance, programmed with rules and with the direct input of human specialists with different domains of expertise, such as microbiology, chemistry, engineering, and geology. These were deployed in diverse industries and fields to support or replace human decision-making and to reduce the cost of human training or moving around human experts in the field.[77] At the end of the 20th century, with the development of big data environments and machine learning, the AI system no longer needed the direct input of human expertise. It would learn and evolve with data and gain autonomy (or semi-autonomy) and agency. Meanwhile, AI systems, such as IBM's Deep Blue, which defeated the human Russian chess grandmaster Garry Kasparov in 1996, were now often presented to the world as evidence of the intelligent computer's increasingly superior powers.

Reflection on the complex human implications of such systems was, of course, not entirely absent in the history of AI. The development of computational autonomous or semi-autonomous systems has always been intertwined with concerns about our human identity and our position in the world. Some of this querying concerns the computer's similarity to humans and, accordingly, the impacts on humans and human societies of these systems that seem to be competing with our human agency and role in society. For instance, if the human neural system is just another information processing system,[78] like the data processing of a machine, why does the human, as just another data processing agent or "inforg",[79] have rights that non-human data processing agents do not have? At the same time, many have also questioned positions such as these by pointing out the fundamental difference between the human mind and consciousness and the mechanical processes of a computer.

In 1980, philosopher John Searle, for example, presented an argument against the idea of human-like AI with his famous "Chinese Room" scenario (see also Chapter 7).[80] It depicts Searle locked in a room, responding to Chinese characters slipped under the door by following a computer programme. Despite not understanding Chinese, he can respond accurately by following the programme's rules, leading those outside the room to believe there is a Chinese speaker inside. This illustrates, Searle argues, that while a computer may be programmed to produce satisfactory responses, it does not mean it has understanding like a human.

Despite very valid questioning, such as Searle's, of the fundamental assumptions of the AI field regarding the similarities between the human brain and computer systems, the understanding of technology as a frictionless extension, or even a direct replication, of human power did not fade. And these ideas also went along well with other impactful technological trends in the late 20th century.

In the history of the Internet and the World Wide Web, early cyber libertarian declarations of independence, for instance, saw any limits on technology innovation as limits to human power. Computers and cyberspace, in particular, were envisioned to become the ultimate extension of human collective power. Internet pioneers and cyber activists such as John Perry Barlow[81] and Donna Harraway[82] imagined a cyberspace that would liberate our human power to challenge and disrupt traditional forms of social power, such as governmental power (and the power of governments to regulate technology innovation) or in Harraway's case the power of the patriarchy. As Barlow stated in his famous Cyberspace manifesto:

> Governments of the Industrial World, you weary giants of flesh and steel, I come from Cyberspace, the new home of Mind. On behalf of the future, I ask you of the past to leave us alone. You are not welcome among us.[83]

Unfortunately, these words of "tech liberation" also inspired the imagination of the early Internet pioneers' "disruption"[84] movement in fundamental ways. "Moving fast and breaking" bounds with existing business practices and markets, also meant breaking with the social rules and prescriptions made by the governments of the offline world. And thus, "tech liberation" also became the main excuse for the rapid development of the most transformative and socially disrupting digital technology services and products of our time. In particular, with the introduction of online social media platforms by Silicon Valley technology industries, privacy and data protection regulations and principles were increasingly presented as inhibiting the evolution of online technology innovation, and the way was pawed for indiscriminate big data business models that track and monitor their users.

Thus, freedom from power was never really achieved. Human power transformed but was not liberated from the restraints of dominant powers. Power did not decentralise as initially imagined by the well-intentioned cyber libertarian human beings but transformed into concentrated clusters of technological power. As argued by a group of experts from MIT, Harvard, Berkley, and other prestigious institutions, the technological imitation of human advantages led to inefficient and harmful technological "centralised architectures" and "centralised power".[85] At the same time, the original ideas invested in AI science and business about the technological "extension of man" evolved into a largely unchallenged narrative about AI superpowers outcompeting humanity. A narrative, they state, that essentially hampers the development of alternative technologies that support humankind "in all its own plurality".[86]

The result of this type of limited imagination about the direct technological extension of human power, one may also argue, was a very simple form of technological innovation and industry. One that has mainly sought to imitate, extend, and essentially replace human power with what Professors Brett Frischmann and Evan Selinger have called "simple machines".[87] In fact, most socio-technical computer systems today have been designed to do this.

Examples of computer systems as such include algorithms deployed to large amounts of personal data to create profiles of humans and act on these as if these profiles were the unambivalent extension of the humans themselves (the "data double" effect[88]); chatbots imitating human feeling and emotion to converse with humans on dating sites; trading algorithms making decisions about buying and selling stocks without the intervention of human stock traders; self-driving cars making decisions about driving in real-time without human drivers; emotion recognition systems analysing human face and voice patterns to detect human emotions making decisions without human intervention in marketing, law enforcement and employment.

The human consequences of this type of "simple" innovation are undisputable. Keith Sonderling, Commissioner of the US Equal Employment Opportunity Commission, a federal agency established in the Civil Rights Act of 1964 to administer and enforce civil rights laws against workplace discrimination, for instance, told me that individuals are left in the dark when autonomous AI algorithms are deployed to make decisions that prevent them from entering the workforce. These decisions are presented without any indication that AI was involved, and it is therefore almost impossible to recognise and challenge the biases and discriminatory policies within companies using these algorithms:[89]

> They may receive a notice that another candidate was selected as more qualified, but they will not be aware that they were passed over due to age, race, socioeconomic background, or another protected characteristic. This lack of knowledge hinders their ability to exercise their civil rights.

The issues extend to job advertisements hidden from some while being in view of others when machine learning techniques are deployed to target only specific groups of people online. In the past, individuals knew job openings existed and had the chance to apply, Sonderling said. Now, these opportunities have become hidden for some:

> We all have a right to the opportunity to enter the workforce, succeed in it, and provide for our family without algorithms making biased decisions. Unfortunately, many of these computer programs merely replicate biases from the past.

"Simple machines" based on limited views of human power are today deployed everywhere, and their implementation in our public and private spheres is growing alongside the evolution of the dominant technology companies of our time. I have provided some examples here of some of these including their human implications. But could we also think of alternative forms of technology innovation?

In 2024, I spoke with the composer, Marianna Filippi (we will return to her a few times in the book). She told me about her creative process when composing music and the way she would sit by her piano for hours to write music in a state of "flow" (see also Chapter 1 on creativity). Then, I asked her if she could imagine an instrument or tool that would work against this creative process. She answered:

> Well, actually, the absence of a keyboard. I always must have a keyboard. I even have a portable keyboard that I use when travelling. Especially when I'm starting a piece, I just need a keyboard everywhere I go, or nothing happens.

The tool is essential; without the instrument, there will be no music, sound, or creative process, Marianna Filippi said. Moreover, it needs to always be with her. However, to meet her needs and empower her human creative process, she also needs a particular instrument, or "technology". Any musician will tell you that their instrument must be fine-tuned to their needs; many will even say they need that *one* instrument with which they feel a special connection. Eric Clapton assembled his famous guitar Blackie from the parts of three different Stratocasters he bought in Nashville.[90] Eddie Van Halen experimented with various guitars and components of guitars to piece together and spray paint his red, black, and white striped guitar Frankenstrat, or "Frankie", to make it sound and play just like he wanted it to.[91] Also, Marianna said she needed that very special instrument to be able to compose her music:

> The piano or keyboard must have some kind of inspiration to it as well. If I haven't imbued myself into the instrument, if I don't get an attachment to the instrument, like if it has no soul, I can do nothing.

Marianna Filippi worried that I found the personal connection she told me she needed with that special instrument weird, but I assured her that nothing is odd when it comes to human relations with their tools – we tend to explain ourselves with mysteries when, in fact, it is just about describing the kind of complexity that humans always struggle with. I then asked her how she knew that an instrument was the one she desired. Her answer was indeed not strange at all. She described in detail the specific technical requirements of the most perfect instrument for a musician:

> It has to do with the technology that I am using. It can be a MIDI keyboard that I can attach to my laptop where I have the sounds of sampled good instruments. If the sounds are too plasticky, it just doesn't do anything for me. For instance, I have this keyboard with a bunch of basic piano sounds and everything, but it just doesn't sound real. There's no soul in it. I can only really be inspired when I'm playing a piano that sounds like a real piano. So, when I attach it to my actual sampled instrument sounds on my computer, it's different. It sounds like a real Steinway, even with the pedal sound.

Marianna Filippi admitted that a real piano, like a Steinway, is her preferred "special instrument". This is why she has the sounds of a Steinway on her computer. A Steinway piano is not a simple tool. On the contrary, it is a very complex instrument that transpires from extraordinary collective human imagination and skillsets. A Steinway is made of Sitka spruce for resonance and hard rock maple for rigidity. The piano keys are made from Bavarian spruce. When the wood for the pianos arrives at the Steinway factory, it is dried for more than two years before it is used. Creating the famous curve of a grand piano takes bending up to 17 layers of hard maple and a mahogany veneer. Then, it is dried again for several months. The strings of the piano also need perfect fitting. Steinway artisans manually nail hundreds of parallel-coded pins to each bridge to guide the strings, ensuring the ideal transmission of sound to the soundboard. Everything must be just so.[92] Marianna Filippi told me that it is just not the same with a digital piano because she needs to "touch" the music and feel the vibration of it. "It's like string theory", she explained, where strings proliferate through space and interact with each other through her body and the piano:

> When it's a real piano, of course, like, oh my gosh, it's endlessly inspirational. It's not just the sound of the piano but also how you feel when you're touching the keys on a real piano and have the lid open; you can actually feel the vibrations, the hammering of the strings and everything. It's a very tangible experience. It's not just like bunking on a couple of pieces of plastic. The reason why I can't get inspired from a fake-sounding piano is because there is nothing tangible about it.

Do you see where I am going with this? Technological innovation does not have to be simple; our tools may be complex and created with their own uniqueness, like the Steinway Piano or Eric Clapton's guitar, which is different from those of the human who commands them. Fundamentally, they should be designed to reinforce human power by providing tools to support, not replace, the complex human potential and capacities.

Fortunately, there are also many computer systems like this. Most of these are not designed for extraordinary, complex tasks, like those of a human composer, but to reinforce human empowerment in our most ordinary everyday lives. These include what computer scientist Ben Shneiderman calls "Human-Centered AI" or human "supertools".[93] He uses real and imagined examples like a "safety first car" (instead of the self-driving car) with a computerised system that predicts risky situations and dangers in traffic to alert the driver; a smart thermostat that does not regulate heating autonomously but supports residents in making decisions with sounds and light responses; spelling and grammar checkers and language translation systems that suggest changes to human writing with delicate red lines; a dishwasher in a dining table; a wheelchair with a joystick or voice control; and an app with exercise instructions.

These are all tools that can empower human beings in their everyday lives. Notwithstanding, also human beings with extraordinary undertakings, can be empowered by "supertools" like this. I spoke with artist Ida Kvetny about AI in her artwork (see also Chapter 1), and when I asked her about AI as a "supertool", she immediately recognised its significance in her life as an artist:

> It's not so much about creating art because I am already really good at that. I don't need help with that. It's all the other things an artist needs to exist. As an artist, you often work alone. For many, it's a one-business show. And there is always this competition with slightly more popular artists who have huge factories, where they have other artists, historians, and assistants who help them with their applications for funding, exhibitions, or other things. Before, I had none. Today, with AI, I have an art studio manager, a curator, someone who can calculate the price of a piece of art for me, write a professional email to a museum, or do the math if I need to scale up something. The many hats that take so much time out of the creative process for an artist. That's when AI becomes a supertool.

AI "supertools" are not stuff for a science fiction movie and perhaps less exciting than the kind of AI that we hear about in the AI-hyped public debate today. They will not outcompete humanity, not take over the world, they will not demand robot rights and not fall in love and then leave us for a more advanced species. In fact, AI "supertools" are not designed and imagined to replace human power. They are computerised systems that work as micro support to humans, leaving the human in control while making human life more manageable. This doesn't mean that we cannot imagine computerised tools like these as critical support for all kinds of extraordinary human tasks performed by not only artists but also medical practitioners, architects, journalists, judges, etc. Indeed, technologies like these can be human superpowers.

In brief, looking at the evolution of technology in the early 21st century, it seems that we have been too rudimentary about what we demand from the tools and instruments we adopt in our everyday lives – for ordinary and extraordinary tasks. Thus, we have much to learn from the musician who requires a special instrument, a special connection with their technology – a physical, intellectual, and emotional interplay that, above all, is respectful of their human needs, their creative human power. The human consequences of these simple technologies are increasingly visible. People are fired by biased algorithms with no human intervention, refused parole in prison, accused of fraud, refused bank loans and health care treatment, receive grades in school, and have their literary, art, musical, and journalistic content stolen and violated by generative AI systems, all due to the adoption of machines that interfere with human powers by disrespectfully imitating and trying to replace it.

Due to their ethical and social implications, we have also had to invest all our political power in governing and regulating the most simple machines that imitate and replace human power, while disregarding our deeply complex human quirks and needs. We simply *had* to put all our political energies into solutions, protections against, and risk mitigation of these simple technologies. It is time now to invest in alternative ways of innovating with the unique human capacities, interests, and needs in mind.

SEVEN TRAITS OF HUMAN POWER

Was the humanistic project conducted in vain? Did we just replace the Almighty God with Algorithmic Gods? Indeed, it seems that in the early 21st century, human power faces significant challenges, when algorithms are unceasingly predicting human behaviour based on past actions, shaping, and directing our lives. And if we continue to embrace this algorithm-driven existence without oversight or understanding the underlying interests and motives, I believe that we are indeed surrendering our destiny and human powers.

Our absolute last stand as humans is a new kind of humanism. One that is not defined against an external other, such as religious dogma, social prescription, or just "technology" but is designed for and with human power. As one of the most influential female politicians in the technology politics of the early 21st century, the Vice President of the European Union and Commissioner for Competition Margrethe Vestager told me in 2023 when asked to reflect on a humanistic approach to technology politics[94]:

> The paradox is that if we want human-centred technology, we need not just to increase our technological speed, we need to increase the speed of interactions between people (…).

A humanistic approach to technology, however, requires a deeper exploration of human power on its own terms. Hence, in the following chapters, we will dig deeper into what this means by delving into seven human traits: creativity, intuition, love, emotion, life, defiance and wisdom.

Why these seven traits specifically? In critical theory, the arts, in literature, film, and in the public debate on digital technology and AI, these traits have always had a special place. Throughout history, they have been scrutinised and depicted, celebrated and praised, but also ridiculed and criticised by artists, filmmakers, authors, gurus, journalists, and scientists to depict the power of humanity as well as our greatest weaknesses.

Why seven? Paradoxically, deploying the number seven to portray human power underscores one of humanity's greatest limitations: our struggle with complexity. Thus, there is really nothing extraordinary about the number seven. It is just an expression of what is averagely human. In fact, if we scrutinise the wealth of human history and culture, we find that we have repeatedly used the number seven to reduce complexity and make sense of our human societies and existence.

In the Hebrew Bible, God created the world in six days and rested on the seventh day. The number seven is mentioned several times in the Quran, such as the seven heavens, seven earths, and seven days of creation. The Christian Bible has seven days of creation, seven sacraments, and seven gifts from the Holy Spirit. There are also in the Roman Catholic faith the seven "heavenly virtues" (humility, charity, chastity, gratitude, temperance, patience, and diligence) to counter the seven "deadly sins" (pride, greed, lust, envy, gluttony, anger, and sloth). In Hinduism, there are seven chakras (energy centres) in the body, and the seven days of the week are named after the seven visible "classical" planets. Think about the requirements for Trustworthy Artificial Intelligence in the European Union's Ethics Guidelines, published in 2019. There are seven: (1) human agency and oversight; (2) technical robustness and safety; (3) privacy and data governance; (4) transparency; (5) diversity, non-discrimination, and fairness; (6) societal and environmental well-being; and (7) accountability. But we know this is not all, don't we?

Numbers play a crucial role in human laws and legal systems by for instance determining the length of sentences, structuring law texts, defining legal thresholds, determining representation or setting rates. In the *Great Binding Law, Gayanashagowa*, the constitutional law of the Iroquois Confederacy comprising the Mohawk, Onondaga, Oneida, Cayuga, Seneca, and Tuscarora peoples, it is stated that "The thickness of their skin shall be seven spans"[95] – meaning that decisions made by the present generation need to be sustainable seven generations into the future.

This is as far as one human generation can see; what they know will happen next to their children and a few generations ahead. Having said that, the idea of seven generations is just one very human way of making sense of and manage the complexity of time and space. Another way could be to imagine the endlessness of it. No human being can see the future. We can consider probabilities, but we cannot truly predict and thus know what will happen next, though we all do intuitively feel – or at least hope – that the future will be there for at least a little more than seven generations. We also have enough knowledge by now to believe this is true. Information has been gathered from studying the soil and rocks of the Earth, as well as the moon and meteorites, to know that the Earth is approximately 4.5 million years old. However, we also know now that the universe is expanding and that it might have an ending. The initial evidence for the universe expanding and possessing a finite age was presented by Edwin Hubble in 1929 when observing with a large telescope the movement of galaxies and calculating the expansion rate. Yet, even though we can't calculate it, and consequently don't know it, could we surpass our human perspective and use our human imaginative powers to see the universe as endless?

Numbers, like the number seven, have meaning when applied in a human context, whether it is laws and religion, or the mathematics used to calculate the expansion of the universe. But they do not present any truths beyond what we know, no evidence to the limits of our human imagination. It is strange, then, that we put so much faith in the numbers and the math behind the simple computer algorithms shaping our human lives today by reducing their complexity.

In fact, in the history of computing, the number seven has played the same role – to reduce complexity for humans – as it did in other contexts. In 1956, psychologist George A. Miller published the paper "The Magical Number Seven, Plus or Minus Two", describing the limits of the human capacity for processing information. What Miller suggested was that the capacity of human memory is limited to around seven (plus or minus two) items, such as digits or words, due to how our brain processes and stores information. "Miller's law" is, for example, applied in computer programming when reducing the complexity of a programme in a structure, also known as "chunking".[96] It has also been used in natural language processing and speech recognition, some of the most well-known AI systems, where information is presented in a manageable way for humans by breaking down complex sentences or text into smaller chunks.

Yes, we have designed the computer in the image of the most limited vision of what the human mind is capable of, not the potential we could imagine it has. We design it to reduce complex inputs and transform it into data that makes sense and can be managed. Take something like "affective computing", a term defined in the 1990s by the scientist Rosalind W. Picard.[97] This is a branch of computer science that studies and develops computer systems that can identify and recognise common patterns ("universal" emotions) into data on individual human emotions. For instance, human facial expressions are coded into discrete categories of emotional expressions.[98] Systems like these could be used for the early detection of autism in children due to their distinctive patterns of facial expressions that are different from other children.[99] But imagine applying this kind of reductive emotional or behavioural identification to a facial recognition system used by the police like the one Mr Williams was exposed to when arrested due to an erroneous match made by the system? Or in a classroom? According to an Article 19 report on emotion recognition systems in China, a market is developing for so-called class care systems. These are systems used in classrooms with children where cameras are continuously taking photos of the class, connecting them to a programme that then uses deep-learning algorithms to detect behaviours ("listening, answering questions, writing, interacting with other students, or sleeping") and give behavioural scores to students that teachers and parents can access via a mobile app.[100]

Today, Rosalind W. Picard warns against the rapid development and adoption of systems like this due to their grave human rights implications.[101] Indeed, affective computing systems' immediate human rights implications are evident when used to control and track citizens. But, another issue, perhaps even more daunting, is the very reductive vision of the human being that these computer systems imply. The first step of an affective computing system is to reduce the complexity of the scale of human feelings and emotions into information that a computer can process. Thus, an affective computing system is essentially not designed to deal with human complexity but to reduce it. As a matter of fact, reduction is the only way the system will work.

Most computer systems work towards reducing complexity. Of course, humans do need to reduce complexity to make it manageable. Sigmund Freud divided the human personality into three components (the id, the ego, and the superego). Karl Marx described three classes of people in society (the bourgeoisie, the landowners, and the proletariat). But we all know that the human mind and society are more complex than that. The computer is an extension of humans' reductive efforts and is undoubtedly helpful in synthesising and reducing complexity here and now. But is it not so that humans have more than just the urge to surpass our deficits by reducing complexity? Do we not also have the imagination and the hope making us capable of creatively constructing a better future for humanity and the planet? Many complex human traits can help us with this, such as creativity, intuition, our feelings and emotions, love, life, defiance, and wisdom. The last thing we want to do is to reduce their complexity and, consequently, our human potential to move beyond simple reduction.

To conclude, like the math behind the computer algorithms, the seven human traits of power that I will explore in the following chapters do not encompass the entirety of human potential, and, of course, that I chose to present only seven of them doesn't reveal any profound or transcendent truths about human nature. Human power is much more complex and could never be reduced to these traits only. I could just as well have chosen to describe 20 or 100 traits of human power, and still, I would never be able to represent human power in all its complexity. We don't know everything about human power, we certainly never will, but we can try to imagine it.

QUESTIONS ABOUT MACHINES

In the following chapters, you should start by envisioning the potential of the seven traits of human power. Then, I'd like you to reflect on these traits in the context of technological development. Throughout the chapters, you will therefore come across standalone questions about machines and AI. Use your human power to reflect on claims about the similarities between technological power and human power. Are they the same? Evidently, they are not. But take a moment and consider why. By doing this, I hope that the descriptions of the seven traits of human power in these chapters will frame your thinking about the power of technology now and in the years to come.

NOTES

1. Talbot, M. (2016, July 17th) "The history of crowd control and the cleveland convention", *The New Yorker*.
2. Tarr, J.A., McShane, C. (2008) "The Horse as an Urban Technology", *Journal of Urban Technology*, 15(1), 5–17.
3. European Parliament. (2000) *Crowd Control Technologies (An Appraisal of Technologies for Political Control)* Final Study, Working document for the STOA Panel, Luxembourg June 2000 https://www.europarl.europa.eu/RegData/etudes/etudes/stoa/2000/168394/DG-4-STOA_ET(2000)168394_EN(PAR02).pdf
4. Beal, A. (2018, May 30th) "In China, Alibaba's data-hungry AI is controlling (and watching) cities", *Wired*, https://www.wired.co.uk/article/alibaba-city-brain-artificial-intelligence-china-kuala-lumpur
5. Information on these systems is derived from an internal study on biometric systems by Aaron Martin and Chris Algrove conducted as part of the InTouchAI.eu project.
6. Podoletz, L. (2023) "We have to talk about emotional AI and crime", *AI & Soc* 38, 1067–1082.
7. See for example: https://neurolytics.ai
8. Angwin, J., Larson, J., Mattu, S., Kirchner, L. (2016, May 23rd) "Machine Bias", *Propublica*, https://www.propublica.org/article/machine-bias-risk-assessments-in-criminal-sentencing
9. Amnesty International (2021, October 25th) *Xenophobic Machines: Discrimination through Unregulated Use of Algorithms in the Dutch Childcare Benefits Scandal*, Index Number EUR 35/4686/2021, https://www.amnesty.org/en/documents/eur35/4686/2021/en/
10. Hasselbalch, G. (2015, May 14th) "Society of the destiny machine and the algorithmic god(s)", www.mediamocracy.org https://mediamocracy.wordpress.com/2015/05/14/society-of-the-destiny-machine-and-the-algorithmic-god-s/

11. Clothilde, G. (2022, September 20th) "Europe edges closer to a ban on facial recognition", *Politico*, https://www.politico.eu/article/europe-edges-closer-to-a-ban-on-facial-recognition/

12. "Senators Markey, Merkley lead colleagues on legislation to ban government use of facial recognition, other biometric technology", press release (2021, June 15th), https://www.markey.senate.gov/news/press-releases/senators-markey-merkley-lead-colleagues-on-legislation-to-ban-government-use-of-facial-recognition-other-biometric-technology

13. "Amnesty International and more than 170 organisations call for a ban on biometric surveillance" (2021, June 7th), https://www.amnesty.org/en/latest/press-release/2021/06/amnesty-international-and-more-than-170-organisations-call-for-a-ban-on-biometric-surveillance

14. Williams, R. (2021, July 15th), "I did nothing wrong. I was arrested anyway", https://www.aclu.org/news/privacy-technology/i-did-nothing-wrong-i-was-arrested-anyway

15. "Another arrest and jail time due to bad facial recognition match" (2020, December 29th), *New York Times.* https://www.nytimes.com/2020/12/29/technology/facial-recognition-misidentify-jail.html

16. Interview 2023

17. "UNESCO in brief", https://www.unesco.org/en/brief

18. Virilio, P. (1989) *War and Cinema: The Logistics of Perception*, Verso.

19. Interview 2023.

20. Thank you Renuka Singh for bringing my attention to "forgetting" and "forgiveness".

21. See the Wikipedia post "right to be forgotten" with reference to court cases worldwide and to the EU's General Data Protection Regulation Article 17 "right to erasure", https://en.wikipedia.org/wiki/Right_to_be_forgotten

22. Council of Europe, Committee on Artificial Intelligence. (2023, January 6th) *Strasbourg, 6 January 2023 CAI (2023) 01 Revised Zero Draft [Framework] Convention on Artificial Intelligence, Human Rights, Democracy and the Rule of Law.* https://rm.coe.int/cai-2023-01-revised-zero-draft-framework-convention-public/1680aa193f

23. Foucault, M. (1991) *Discipline and Punish: The Birth of a Prison.* Penguin (originally published in French in 1975).

24. Foucault (1991/1975).

25. Penta, L. J. (1996) "Hannah Arendt: on power *The Journal of Speculative Philosophy*", New Series, 10(3), 210–229.

26. Arendt, H. (2018) *The Human Condition*, 2nd ed. (p. 11), Chicago University Press (originally published in 1958).

27. Ball, Stephen J. (1993) *An Horizon of Freedom: Using Foucault to Think Differently about Education and Learning*, Routledge.

28. Rizvi, J. (2019, June 30th) "Open-Plan Work Spaces Lower Productivity And Employee Morale", *Forbes*, https://www.forbes.com/sites/jiawertz/2019/06/30/open-plan-work-spaces-lower-productivity-employee-morale/#46528a461cda

29. Berstein E., Waber, B. (2019, November–December) "The truth about open offices", *Harvard Business Review*, https://hbr.org/2019/11/the-truth-about-open-offices

30. Cain, S. (2012) *Quiet The Power of Introverts in a World that Can't Stop Talking*, Penguin.

31. Cain (2012, p. 75).

32. Cheng, J. (2012, June 1st) "The Slow Web", https://www.jackcheng.com/the-slow-web/

33. Hasselbalch, G. (2021) *Data Ethics of Power – A Human Approach in the Big Data and AI Era* (p. 165), Edward Elgar.

34. Ucnik, L. (2022) "Hannah Arendt's action and contemplation: two sides of the same coin", *Journal of Social Philosophy* 53(1), 76–92.

35. Ucnik (2022, p. 83).

36. White, T. "What Hannah Arendt really mean by the banality of evil?" *Aeon*, https://aeon.co/ideas/what-did-hannah-arendt-really-mean-by-the-banality-of-evil

37. Ucnik (2022).

38. Smuha, N. A. (2022) *The Human Condition in An Algorithmized World: A Critique through the Lens of 20th-Century Jewish Thinkers and the Concepts of Rationality, Alterity and History*, Institute of Philosophy, KU Leuven.

39. Cohen, J. E. (2013) "What privacy is for", *Harvard Law Review*, 126(7).

40. Stoycheff, E. (2016) "Under surveillance: examining Facebook's spiral of silence effects in the wake of NSA internet monitoring", *Journalism & Mass Communication Quarterly*.

41. Hasselbalch, G., Tranberg, P. (2016) *Data Ethics. The New Competitive Advantage*, Publishare.

42. Interview 2023.

43. See: https://recoil-performance.org/productions/act-of-gravity/

44. Interview 2023.

45. Cain (2012).

46. Arendt (2018/1958, p. 11).

47. Penta (1996, p. 212).

48. Bergson, H. (1977) *Two Sources of Morality and Religion* (translated by A. Audra & C. Brereton), University of Notre Dame Press (originally published in French, 1932).

49. Said about complementarity in music at Resonans festival – 20th august 2022 (in Danish: "Man lytter sig ind til hinanden").

50. Baudrillard, J. (1988) "Simulacra and simulations" in Mark Poster (ed.), *Jean Baudrillard, Selected Writings* (pp. 166–184), Stanford University Press.

51. Hasselbalch (2015, 2021).

52. Hasselbalch (2021).

53. Guo, E., Renaldi, A. (2022, April 6th) "Deception, exploited workers, and cash handouts: how Worldcoin recruited its first half a million test users", https://www.technologyreview.com/2022/04/06/1048981/worldcoin-cryptocurrency-biometrics-web3/

54. O'Leary, L. (2022, January 3rd) "how IBM's Watson went from the future of health care to sold off for parts", *Slate*, https://slate.com/technology/2022/01/ibm-watson-health-failure-artificial-intelligence.html

55. Edwards, J. (2024, June 20th) "Poll reveals Americans' fears about AI", *NewsWeek*, https://www.newsweek.com/poll-reveals-fears-ai-smarter-attack-humanity-1915100

56. Sartre, J. P. (1989) "Existentialism is a humanism" (translated by Philip Mairet) in Walter Kaufman (ed.), *Existentialism from Dostoyevsky to Sartre*, Meridian Publishing Company (originally in French 1946), https://www.marxists.org/reference/archive/sartre/works/exist/sartre.htm

57. Schröder, S. (2021) "Humanism in Europe", in Anthony B. Pinn (ed.), *The Oxford Handbook of Humanism*, *Oxford Handbooks Series* (online ed., Oxford Academic, 4 Oct. 2019).

58. Schröder (2019).

59. The demonstrations at the Chicago Democratic Party convention held on August 26–29th, 1968, led to violent clashes between young anti-war protesters and police in the streets of Chicago, where five of the protesters were later given prison sentences and fines. See the CNN's depiction of the events: https://edition.cnn.com/ALLPOLITICS/1996/conventions/chicago/facts/chicago68/index.shtml; the trial of seven of the protesters is depicted in the Netflix movie *The Trial of the Chicago 7* (2020).

60. See for example film critic of the Chicago Sun-Times and Pulitzer price winner Roger Ebert's review of the movie in 1970, who among others wrote: "This is such a silly and stupid movie, all burdened down with ideological luggage it clearly doesn't understand, that our immediate reaction is pity." https://www.rogerebert.com/reviews/zabriskie-point-1970

61. Bergson, H. (1914) *Creative Evolution* (translated by Arthur Mitchell) (p. viii), Macmillan and Co. (Originally published 1907)

62. Kaiser, D., McCray, W. P. (eds.) (2016) *Groovy Science Knowledge, Innovation, and American Counterculture*, University of Chicago Press.

63. Moreno, J. D. (2014) *Impromptu Man J. L. Moreno and the Origins of Psychodrama, Encounter Culture, and the Social Network*, Bellevue Literary Press.

64. Moreno (2014).

65. Richard, B. (2015) *Jonathan Livingston Seagull*, Harper Thorsons (originally published 1970).

66. Edwards, P. (2002) "Infrastructure and modernity: scales of force, time, and social organization in the history of sociotechnical systems", in T. J. Misa, P. Brey, Feenberg, A. (eds.), *Modernity and Technology* (pp. 185–225), MIT Press.

67. Lyotard, J. F. (1984) *The Postmodern Condition: A Report on Knowledge.* (translated by Geoff Bennington & Brian Massumi) (p. 4), University of Minnesota Press (originally published in French in 1979).

68. Lyotard (1984/1979, p. 17).

69. Rand, E. K. (1932, Apr.) "The humanism of Cicero", *Proceedings of the American Philosophical Society* 71(4), 207–216.

70. Cicero, "On divination" 1.125–6, trans. Long and Sedley 1987, 55L, from Keith Seddon (1999) http://people.wku.edu/jan.garrett/stoa/seddon1.htm
71. See also Lapenta, F. *Our Common AI Future – A Geopolitical Analysis and Road Map, for AI Driven Sustainable Development, Science and Data Diplomacy*, JCU Future and Innovation Publishing, https://dataethics.eu/our-common-ai-future/
72. World Economic Forum, "Davos23: five tech superpowers transforming humanity and technology", *World Economic Forum Agenda, 2023*, https://www.weforum.org/agenda/2023/01/davos23-five-tech-superpowers-humanity-technology/
73. McLuhan, M. (2013) *Understanding Media: The Extensions of Man*, Gingko Press (originally published 1964).
74. McLuhan (2013/1964).
75. Moor, J. (2006) "The Dartmouth College artificial intelligence conference: the next fifty years", *AI Magazine*, 27(4), 87–91.
76. Goel, A. (2022) "Looking back, looking ahead: humans, ethics, and AI", *AI Magazine*, 43(2), 267–269.
77. Crevier, D. (1993) *AI: The Tumultuous History of the Search for Artificial Intelligence*, Basic Books.
78. Wiener, N. (2013) *Cybernetics or, Control and Communication in the Animal and the Machine*, 2nd ed., Martino Publishing (originally published 1948); Bynum, T. (2010) "The historical roots of information and computer ethics", in F. Floridi (ed.), *Information and Computer Ethics*, Cambridge University Press.
79. Floridi, L. (1999) *Philosophy and Computing: An Introduction*, Routledge.
80. Searle, J. R. (1980) "Minds, brains, and programs", *Behavioral and Brain Sciences*, 3(3), 417–457.
81. Barlow, J. P. (February 8th, 1996). "A declaration of independence of cyberspace", https://www.eff.org/cyberspace-independence
82. Haraway, D. (1991) "A cyborg manifesto: science, technology, and socialist-feminism in the late twentieth century" in *Simians, Cyborgs and Women: The Reinvention of Nature* (pp. 149–181), Routledge.
83. Barlow (1996).
84. See Christensen, C. M. (1997) *The Innovator's Dilemma: When New Technologies Cause Great Firms to Fail*, Harvard Business School Press. He describes business practices that successfully innovate by breaking with existing practices and markets and creating new ones.
85. Siddarth, D., et al. (2021) "How AI fails us", Justice, Health, And Democracy Impact Initiative & Carr Center for Human Rights Policy, https://ethics.harvard.edu/files/center-for-ethics/files/aifailsus.jhdcarr_final_2.pdf?m=1651510742
86. Siddarth (2021).
87. Frischmann, B., Selinger, E. (2018) *Re-Engineering Humanity*, Cambridge University Press.
88. Haggerty, K.D., Ericson, R. V. (2000) "The surveillance assemblage". *British Journal of Sociology*, 51(4), 605–622.

89. Interview 2023.
90. See https://whereseric.com/faq/blackie-eric-claptons-fender-stratocaster/
91. See https://www.guitarworld.com/features/eddie-van-halen-frankenstein-origins
92. See https://www.steinway.com/
93. Shneiderman, B. (2022) *Human-centered AI*, Oxford University Press.
94. Interview 2023.
95. Constitution of the Iroquois Nations, "The Great Binding Law, Gayanashagowa", https://cscie12.dce.harvard.edu/ssi/iroquois/simple/1.shtml
96. Gee, S. (2020, March 19th) "The magic number seven and the art of programming", https://www.i-programmer.info/babbages-bag/621-the-magic-number-seven.html
97. Picard, R. W. (2000) *Affective Computing*, First, Massachusetts Institute of Technology (originally published in 1997).
98. Picard (2000/1997, p. 176).
99. Talaat, F.M. (2023) "Real-time facial emotion recognition system among children with autism based on deep learning and IoT". *Neural Computer & Application* 35, 12717–12728.
100. Article 19 (2021) *Emotional Entanglement: China's emotion recognition market and its implications for human rights* (p. 27). https://www.article19.org/wp-content/uploads/2021/01/ER-Tech-China-Report.pdf
101. Picard, R. (2017, June) "Affective computing, emotion, privacy, and health, artificial intelligence podcast by Lex Fridman", https://www.media.mit.edu/articles/rosalind-picard-affective-computing-emotion-privacy-and-health-artificial-intelligence-podcast/

CHAPTER 1

Creativity

Does a Machine Have a Creative Impulse?

One day, the Society of Independent Artists' salon in New York received an odd piece of art. Or maybe it wasn't art after all? It was a very mundane object; the most creative thing about it was how the artist had positioned it. Upside down. The inverted porcelain urinal, because that is what had been submitted to the exhibition, was signed by the unknown artist, R. Mutt. He had also given it a much prettier name than such a thing would typically deserve: "Fountain". It was April 1917, and as a rule, the Society board of this open avant-garde art establishment would accept any piece of art for a fee. All the same, this piece most of the board didn't want to include in the exhibition: a urinal is not an actual work of art just because an artist signs it. Except for one board member, Marcel Duchamp. He held that this upside-down urinal was, in fact, an artistic expression of creativity. And, of course, he did because R. Mutt was a cover name for Duchamp himself, who was the one who had submitted the Fountain. His creativity was expressed in what is well known today in the art world as a "ready-made", an everyday thing or product repurposed as art, challenging the set notions about the sources of artistic creativity and the culture it represents.

The refusal to exhibit a urinal at an avant-garde, supposedly open-to-all art exhibition, created an essential debate at the time: What is creativity? Where does it emerge? Is it the origin of an authentic human creation,

DOI: 10.1201/9781003527855-3

a painting or a sculpture created single-handedly by an artist? Or is it a process emerging in the interface between a human and an object, tradition, and innovation? To whom or what does creativity belong? To the one or the many?

Does a machine have a creative impulse?

The kind of creativity we see in cultural production is often creation that differs, is distinct, new, and surprising. The most creative painter paints an extraordinary painting. The creative filmmaker will make a movie that seems different from what we've seen before. The creative app developer will create a new app that meets a need you didn't know you had. All these people will use existing materials, ideas, and tools, but something new emerges from the process.

Creating new things is also what generative AI does, and at the moment, we are surprised every day with new AI-generated pictures, videos, or text that seem like something from tomorrow. Dogs that climb on walls, pineapple heads, new songs from long-gone musicians. A generative AI system can create an image that has never been seen before based on the millions of humanly created images from the web it was trained on. While it is easy to confuse these new creations with human creativity and creative processes, let's reflect in more detail on how human creativity differs from the data processing of generative AI.

First, human creativity goes beyond the mere creation of the novel. For instance, as I will illustrate by tracing ideas about human creativity in this chapter, it is a force of life and an interplay between this human life and the cultures and societies in which we exist. In these views, creativity springs from human experience and feeling; it is "fun" and unpredictable, an openness to not knowing what comes next, and a direct experience of time and human history. Creativity often emerges in critical moments where established ideas clash and controversy arises. It feels like a human urge or a necessity. Rarely does it make immediate sense in the context of established cultural norms and rules in which it exists. And all this human creativity is, in fact, a fundamental pillar of modern economies.

Let's begin with the cultural theorist Raymond William's critique of elitist ideas of what constitutes authentic culture. I believe this provides part of the answer to the questions posed by Duchamp in 1917 about human creativity. Williams emphasises cultural meaning production as something that emerges between what we know and have "been trained to" and "new observations and meanings, which are offered and tested".[1]

This is why, as he says, culture is not just that which we find in a museum and have been accustomed to; it is multiple, "a whole way of life". Culture is both "traditional" and "creative", meaning that it is, as he also states, "made and remade in every individual mind".[2] To Williams, creativity is, therefore, first and foremost, this open human interpretation and challenge of cultural tradition. It is the mobile side of culture, that which is not rooted and stable. It is movement, mobility, and openness, a process that is not steady and predictable.

Cultural theorists, such as Williams, describe culture as meaning-making systems that sustain human communities that share ideas, priorities, and goals. Culture is, therefore, institutionalised, formalised, and practised by dominant groups in society. However, it is also always up for contestation and social negotiation. The way we value and reward human creativity in society is always in the making. The most creative human being will manage to comprehend and grasp the dichotomies of human culture and transform them into something meaningful. And so often, human creative expression in art, for instance, is most interesting when it manages to grasp human culture with all its values, rules, and norms about what is beautiful and ugly, good and bad, and right and wrong while at the same time challenging these things altogether. In this way, human creativity also exists in cultural spaces that are at once stable and in the making.

How is culture expressed in AI training data? Can AI challenge the data it has learned from and is trained on?

When Duchamp submitted the Fountain to the art exhibition, the upside-down urinal was this: open to interpretation, challenging the stable and fixed meaning of the art establishment and its static notions of what constitutes and is valued as art and creativity in a specific human society and culture. Today, more than 100 years later, the "ready-made" is a cultural tradition. We know what it means and how to appreciate and approach a piece of art like that and we understand how to interpret the human creativity invested in a "ready-made".

Many things have since then been repurposed to create conceptual art, where the artist places an object in a new context to generate meaning and provoke us to think differently. With the "Lobster Telephone", Salvador Dalí in 1936 combined a telephone with a lobster. Two juxtaposed seemingly unrelated objects; even so, in Dali's surrealist art, the lobster had always had strong sexual associations that were then channelled through the phone in a time when the "sexual" was repressed and unspoken of.[3] In 1998, in the art installation "My Bed", Tracey Emin presented her unmade bed with vodka bottles, cigarette buds, dirty socks and underwear, bringing us into the intimate and exposed sphere of the artist. Jeff Koons' "Balloon Dogs", created between 1994 and 2000, is a series of massive, shiny, stainless-steel sculptures shaped like balloon animals. Koons takes these ordinary objects from a child's birthday party and blows them out of proportion. Known to have rejected any grand interpretation of these shiny objects, his refusal to present a fixed meaning is perhaps precisely the meaning of these "ready-mades".[4]

Human creativity doesn't lend itself easily to a fixed interpretation, as we will see later. It often escapes it. This is the strength of human creativity. It changes. It is more of a fleeting movement in time than a fixation of human thought and practice in space. I will get back to why this is so and how some see this feature as the core strength of human creativity.

Is OpenAI an artist of the "ready-made"? Is DALL-E a "ready-made" piece of art?

In the summer of 2024, I visited artist Ida Kvetny's studio in Copenhagen to talk with her about her creativity and work as an artist. She is an interdisciplinary artist working across various media, incorporating AI, AR, VR, and classical art forms into her practice. In her work, she uses, for instance, virtual reality to combine digital elements with her traditional works in paint and clay. I had met her before at an event at the Museum of Modern Art in Copenhagen, where we were speaking at an 8th of March International Women's Day event on women, art and algorithms. Back then, I was bemused by her digital worlds on display with huge, deformed, human-like figures walking awkwardly across imaginary landscapes of splendid colours and textures. At the same event, she transformed the traditional museum statues of male figures into female-like virtual avatars when the visitors pointed their phones at them. In the studio I was now visiting, huge canvasses were covering the back walls, and ceramics bent out of ordinary shape were spread across the floor – here, a limb sticking out of a vase; there, an eye staring back at me. Repeating the universe and colours of the virtual realms I had seen a few years before at the museum.

As an artist, Ida Kvetny sees her surroundings and everything in them as spirited, she told me. "Think of the world", she explained, "as something that is not made up of just boring things, but use your imagination to enchant them, interact with them and talk with them. You can do that at different levels", she said, "and when doing so, you make the world more imaginative … or human".

When generative AI, such as the image and video-generating AI models, DALL-E, Stable Diffusion and Midjourney, became world news and people across the world started testing them and falling in awe over their "creative" capabilities when generating extraordinary content based on prompts, Ida Kvetny had already for a long time been using digital tools and AI in her artistic practice. She told me that she has always enjoyed working in "co-creation" sessions, whether working with a musician or a writer and as a cross-media artist, she appreciates using one medium to describe another, for instance, finding ways to explain things differently and see them from a new perspective. She believes in being open to other interpretations of her art, daring to step back and relinquish control. For her, being creative has thus always been a balance between the very controlled and the uncontrolled. This is also why she immediately found it interesting to work with AI. The AI model joins in, co-determines and makes some choices, and you lose some control, she explained, but at the same time, you don't have to lose all of it. It's like a white canvas, where you paint with a brush together with another brush that is not your own, she said:

> It's a good way to jam with the imagination, but it's also a way to examine a kind of human subconscious, our collective consciousness. Generative AI is like a mirror image in the water. A kind of echo that you can resonate with. You have a character, and then you ask the algorithm to speculate on what happened in the character's story. And when it does, you can see what these algorithms are trained on. Why, for example, does this sculpture keep turning into a David figure? Or why does it always remain only this Greek, classical god? You can also stick a finger in that mirror image in the water and make it dissolve.

Ida Kvetny's studio is at the bottom of a building that has been housing artists of all generations on each floor since the 1980s. You first enter a huge bright room through one façade of window glass. This is where she paints on large canvasses or works by the big table with the grand computer screen. Where her oddly shaped colourful figures hang from the ceiling. But, when the freezing cold enters through the large windows in the room in the winter, she withdraws to the womb of the studio, a small windowless space shielded from the rest behind large, heavy curtains. To Ida Kvetny, art is precisely that: a space, but not just one where you can paint on large canvases or programme virtual four-dimensional worlds and reach exorbitant dimensions. It is a critical, creative space in time:

> I see art as a space, where you can discuss some things. A poetic place where you can turn things upside down. It doesn't need to end up in a product or a finished thing. It is an experimental place where you collaborate and communicate. Whether it is with brushes or with words or with other things. But that space, it is so essential, that is what makes us human.

In my last book, I wanted to describe special human moments in history and society where new and old cultural ideas and perspectives emerge, such as those we have on our human creativity, and where existing ideas and their norms are challenged. These "critical cultural moments",[5] as I called them, are "ephemeral, they do not last".[6] Usually, they are absorbed into the dominant culture when institutionalised and turned into technical design, for instance, or laws and rules. They have their own special time, and then they are gone. Ida Kvetny told me that she felt a bit like that about AI and art. It had a moment, and it was this moment she liked to depict. She explained:

> What I have found interesting in my work with AI was documenting the transition in it. For example, if you, at the beginning, when the first generative AI models came out, wrote a prompt about a horse made from mushrooms in a field, then you got this strange horse made from mushroom clouds. But after two generations of the model, you could write the same prompt, and then you got a boring horse standing in a perfect mushroom field. The models were fixed.

Ida Kvetny told me that she didn't want to judge whether the development she was documenting was good or bad. But she does see a challenge in the fact that AI, in many ways, streamlines, smooths out and cleans up the human world, which is, however, also, as she said, in essence, a very human thing to do:

> That's how it is; we humans are constantly trying to fix things: all those little cracks and all these little mistakes. Always must be fixed and adjusted. But that's where the poetry is. It's been interesting to see how something starts, and then it's a bit helpless because, of course, it's humans who create it, which means there are mistakes. There is bias, like racism, in it, all these things that we would like to correct. After all, we want to create better societies and try to improve things for people, right? Somehow, it is also interesting that AI shows who we are, for better or for worse.

The AI "critical cultural moment" seemed to have passed for Ida Kvetny, and AI was to her becoming more like the "supertool"[7] described in part 1 of this book. Less of a creative "co-designer" that sparked her creative impulse with odd responses to her prompts and more of an assistant for "boring tasks", such as calculations and email writing.

Critical cultural moments are indeed sensitive, ephemeral at the least, and they are certainly not a given. This is because, as I described in my last book, they have "special human characteristics and are possible when human memory and intuition are privileged and provided time and space to tinker",[8] something that is not always prioritised in society, as I have tried to illustrate in the first part of this book. And so, when the moment where the co-creation process sparks the human creative mind to wonder about the machine's "mistakes" and "quirks" is no longer there, when the algorithm is fixed, when there is no longer a clash between the human prompt and what the machine can come up with, when the generative AI tool creates a perfectly looking horse on a perfect field making perfect sense, as Ida Kvetny explained it to me, that is, one could argue, also when the creativity stops.

The "critical cultural moments" I also argued require "spaces of negotiation" that enable critique and negotiation. These spaces are only possible when various systems clash and controversy arises.[9] Here, I wanted to describe the moments in technology politics where ethical discussions emerge among decision-makers, but not much is certain. And perhaps it is precisely this "creative space" that the artists excel at exactly because this is the space where they feel most "at home"? Once, an artist who had read my book came to me at an event and told me that as an artist, he knew so well what I was trying to describe with these two concepts, and Ida Kvetny was confirming with me what I was suspecting. Certainly, we should not fear that new generative AI models will replace the artist as all they do is produce; they do not create anything in critical cultural moments as the artist does for a living or as the creative decision-maker does when navigating and negotiating uncertainties just before a law is adopted. An artist's creative process is a human critical space rather than an end result – a painting, a sculpture or a video. The creative process of a decision-maker is a human critical space rather than an end result – a law or a policy document. This is also why Ida Kvetny does not feel endangered as an artist by the arrival of a new "generative" tool. As she says:

> After all, art has also changed over time. It is not stagnant. It develops with our tools and our society. It is like the arrival of photography or the printing press. People were also scared that they would replace the human artist. But they have just become part of the umbrella of possible tools.

Artists are not like we imagine the "traditional artist" anymore, she continued. Their creativity revolves around the idea of combining the oddest kinds of expertise and making connections:

> What we fear is disappearing already disappeared with Duchamp's urinal and Andy Warhol's soup cans. It's all about the idea. Most artists work with concepts. They are already combining something or have some technicians to build something for them. So, whether it's a stick that you sit and draw in the sand with or it's artificial intelligence that you work with to transform that idea doesn't really matter.

As an artist experiencing the AI moment in the art world and beyond, she even sees an opportunity to do what artists have always excelled at, which is to use their creativity to explore the world by asking questions:

> Artists have a really strong investigative apparatus. That's what artists do. We ask all the time, why am I doing it? What am I going to use it for? So, what we can do with the question of AI is that we can interact with these models. We can ask some strange questions and we can make some strange connections. After all, our generation has a huge obligation. We are a transitional generation. These models are built on our data, images, and conversations. And it's pretty crazy to have the responsibility as an artist of that generation, to have to comment on that with your art.

When in 2004 journalist Ed Bradley asked sing song writer Bob Dylan about the origin of the song "Blowing in the Wind", he responded that it came from a "wellspring of creativity".[10] Often, human creativity is described as something that "special human beings" like Bob Dylan or Ida Kvetny possess: artists, filmmakers and authors. But let's also try to understand human creativity as a fundamental feature of being human. Some humans, like the artist Ida Kvetny, might indeed be better at harnessing this human power when creating extraordinary imaginary worlds on canvas or in virtual reality. However, it can also be said that creativity is a key feature of human life in general, a creative impulse that propels life forward.

Henri Bergson famously depicted the "creative evolution", a type of evolution of a species that is not conditioned by the past, like that of Charles Darwin's mechanisms of natural selection and adaptation. Bergson describes evolution as constantly creating and recreating itself, changing the whole continuously and thus always moving towards an undefined future. "(…) to exist is to change, to change is to mature, to mature is to go on creating oneself endlessly", as he wrote in 1907 in the introduction to his book *Creative Evolution*.[11]

Perhaps this is precisely the kind of creative evolution of life that David Bowie pondered about when he, at his 50th birthday concert in 1997 in Madison Square Garden, exclaimed:

> I don't know where I am going from here, but I promise I won't bore you….

David Bowie had been moving confidently through life with a creative drive that fundamentally challenged in music and style the established norms of the music industry and our ideas about what you are supposed to look and be like and yet hadn't found his direction. His human creativity was not bound by any norms, defined in advance, but much more like Bergson's creative life force, unpredictable, free, open, and unconditioned.

Henri Bergson considered the open perceptive experience of life in the making, of time, or "duration", as he preferred to call it, most faithful to reality as is. He argued that representing reality in immobile data, numbers, and narrative, "spatialising time", might be useful in the natural sciences. Nonetheless, it does not help us to understand our true nature, which is always in the making, unconditioned, and unpredictable. The artist, on the other hand, he thought, has a unique ability to perceive duration,[12] "to think movement".[13]

In essence, human creativity is a type of curiosity that does not preoccupy itself with "truth", as this would defeat the core impulse of creativity to engage with life and reality in the making. As one of the most creative philosophers of our time, Slavoj Zizek, says:

> Truth is neither "subjective" nor "objective": it designates simultaneously our active engagement in and our ecstatic openness to the world, letting things come forth in their essence.[14]

David Bowie didn't know what the future would hold for him, but he knew it would be fun. The creative human impulse is playful, taking risks and curious, seeking the unknown rather than what has been done before. As film school Director Bo Erik Hasselbalch once told me: "It is the unknown that gives the known quality".

This is also how the psychologist Mihaly Csikszentmihalyi describes the creativity of the humans he has studied. These humans proved to be particularly creative as scientists, politicians, and artists.[15] He illustrated how they love what they do; they don't just do what they do because of the money or because they must.

Does a machine have fun, take risks, and seek the unknown?

Creative humans take joy in their creativity. This doesn't mean that acting on one's creativity is not an effort, sometimes immensely so. It is hard work to act on a creative impulse and make something of it, whether it is a piece of art, a political statement, a different way of approaching a scientific topic, designing a new app, or writing a book about human power. I have isolated myself in a little house by the sea while writing these words about human creativity. My brain aches, and my back hurts from sitting in the same position in front of the computer for hours. I force myself to finish a section before I allow myself to go for a walk. If I don't continue, I will lose the flow of the creativity I need to finish this.

Csikszentmihalyi described many psychological features of human creativity, but most famously, he explored the creative human mind as one in a state of flow. He met artists who would become immersed and disappear in their work, unaware of their surroundings or essential needs like food and sleep. They would describe this state of mind as water flowing with the currents.

The Italian-American composer Marianna Filippi described to me her state of flow as the moment in her creative process when everything comes together: emotions, ideas, inspirations, intuitions, instruments, and musicians and thus where she gets the feeling that everything is attainable:[16]

> (...) the thing that I'm trying to do is right there. Like I can see it. I can feel it; I can hear it. It's like you're finally caught up with this unattainable moment. You're finally there and unite with it. It's amazing when that happens because I know exactly what I want. I know exactly what I'm doing, what I want to write, and what instruments I want to use. I know exactly what kind of theme I want. It's a feeling of just knowing and not having to be chasing it. You know what you want immediately.

This moment when everything comes together and is attainable is when she forgets everything around her and gets into a state of flow. She recalls last summer when she was writing her previous symphony:

> I just sat there for like 12 hours a day, just writing and writing and writing because I knew exactly what I wanted.

Filippi finds it hard to put words to this moment. She can't tell where it is coming from, and it feels like a powerful impulse from inside that she must submit to.

> (...) When you're in the flow state, things happen exactly as you wanted it to happen. And there's no forcing it. There's no trying to spark your mind to keep going. It just happens. It's a very rewarding experience because you realise you knew what you wanted this whole time. Like, I'm not questioning anything. I'm just going with it. And it's submitting yourself to this, to the potential fear, I guess, of the unknown and going with it. It is honestly the old saying, "Just go with the flow".

Flow is the closest we can get to Henri Bergson's idea of "thinking movement", which is to experience time and duration directly. In other words, a human being's creativity can be described as a direct and uninterrupted experience of time and place. As such, it is hard to represent in any recognisable form as the moment you express it, it is no longer creativity. What is left of the original human creativity in a piece of art at an exhibition, for example, is what philosopher and literary critic Walther Benjamin called its "aura". The unique trace of human time: the paper that has yellowed over time, the pencil stroke, the interrupted line, the meeting between the creator, the creation and the receivers in the time of its making. In Ida Kvetny's studio, we stood by a small table, turning some 3D-printed sculptures in our hands. She pointed out the lines in the plastic and the many little strings of plastic sticking out of the sculptures that were not supposed to be there. These she particularly liked, she said, as they were the traces of the machine they were made with. While we were standing there, I noticed she was picking at the strings. Leaving some on while pulling some out. She was leaving her unique trace on the sculpture. The "aura" of art is impossible to replicate, as Benjamin wrote in his famous 1935 essay "The Work of Art in the Age of Mechanical Reproduction":

> Even the most perfect reproduction of a work of art is lacking in one element: its presence in time and space, its unique existence at the place where it happens to be.[17]

Can a machine reproduce the human aura?

Creativity emanates from lived experience, human emotion and feeling. In 2022, after the OpenAI language model ChatGPT was launched to the public, excited fans of the poet and musician Nick Cave started sending him AI-generated lyrics based on prompts to write in his music and poetry style. However, Cave did not see the resemblance; in fact, he rejected even the slightest likeness in its entirety. A machine like ChatGPT, as he said, cannot write a genuine song as it lacks authentic human experience:

> Songs arise out of suffering, by which I mean they are predicated upon the complex, internal human struggle of creation and, well, as far as I know, algorithms don't feel. Data doesn't suffer. ChatGPT has no inner being; it has been nowhere, it has endured nothing (...).

Creativity can only be human, according to Cave, as creativity stems from human lived, felt experience and from human struggle. In his case, he is referring to the most painful human experiences, which he has wrenched into the lyrics of his songs over the past years. He had lost two of his four sons. One just the year before. This very personal shape of creation doesn't mean that the creativity that his lyrics stem from is only his; all it means is that it is an intensely human experience that may become intertwined with another human being's experience when listening to his songs:

> This is part of the authentic creative struggle that precedes the invention of a unique lyric of actual value; it is the breathless confrontation with one's vulnerability, one's perilousness, one's smallness, pitted against a sense of sudden shocking discovery; it is the redemptive artistic act that stirs the heart of the listener, where the listener recognizes in the inner workings of the song their own blood, their own struggle, their own suffering.[18]

Can AI perceive time? Are the futures that it predicts open? Are they unconditioned by the past?

The richness of the unique human experience is transformed into a creative impulse. It is expressed not only in content but also in the form and style of a creation. If we take our time and look carefully, we may see the creativity of human experience expressed in the tiniest of details.

Karen Blixen lived an extraordinary life, which became the foundation of her creativity as an author and infamous storyteller. She was born in 1885 and grew up with her wealthy Danish family in their estate, Rungstedlund, on the seaside in a town not far from Copenhagen. In 1914, Karen Blixen married her cousin, Baron Bror Blixen-Finecke, and moved to Kenya in Africa to establish a coffee plantation. Although they divorced in 1925, Blixen decided to remain in Africa and run the coffee farm on her own. She was fascinated by the people and the land, and she wrote captivating stories and memoirs about her time in Africa, among others, under the pseudonym Isak Dinesen; she published Out of Africa in 1937.

We see Karen Blixen's rich human life depicted directly in her stories. We can also read about or watch the countless depictions of her life made by other authors or filmmakers. Having said that, Blixen's creativity is most profoundly expressed in the details of her language. Like only a few other authors, such as Samuel Beckett, Milan Kundera and Joseph Brodsky, she insisted on translating her work from Danish to English. After a while, she started writing and narrating her stories orally in English.[19] The English language stories that made her world-famous were characterised by her "Danicisms" and often Danish odd sentence structures translated into English. At the same time, her interest in her African home and its people led her to practice Swahili. The combination of these languages in her stories expressed her human experience and identity as a foreigner, always in between places. She no longer had just one language in which to express herself. Blixen told her stories in a "hybrid" language representing her sense of "foreignness", as professor in literature Lasse Horne Kjældgaard describes it:

> (…) there was no pure and natural language to fall back on but rather a hybrid language, a borrowed language. In Africa, she had been a foreigner but she was also foreign when she came "home" to Rungstedlund. The body of writing she created and which still appeals, even today, to thousands of readers all over the world, is based largely on this experience of foreignness.[20]

What is the cultural identity and style of a machine?

We see creativity expressed in the work processes and lives of the most artists, scientists, authors, enriching human science and cultures. We feel creativity as a fundamental life force. Is it not evident then that human creativity is also a source of social and economic prosperity?

The urban studies and economy professor Richard Florida[21] has famously described the human creativity of an entire class of people with their histories, preferences, and shared modes of thinking and working as an economic force. He says that this "creative class" of people is as crucial for modern-day economies as natural resources, physical labour and capital was for the economy of the industrial age. The creative class is rising, particularly in the world's metropolises. It consists of people who make a living from creative work and contribute to an economy that progressively depends on their skills. They have technological, economic, artistic, and cultural creativity, share similar "thought processes", and tend to gather in specific environments that meet the fundamental properties of creativity, such as diversity, openness, and mobility.

Florida argues that failing to cater to this kind of human creativity can restrict its potential and limit its growth. Therefore, the key to economic growth in a contemporary city depends, as he says, on its openness to all forms of creativity. Embracing diverse creative expressions is fundamental to attracting the creative class, ensuring that the city evolves and thrives as a hub for innovation and talent. It also depends on economic means. An environment that economically invests in human creativity and creative production will also see an upsurge in creative industries.

An OECD book from 2022 recognises that investing in cultural and creative sectors (CCS) benefits economic and social development. It is not only a significant source of creative output, jobs, income, and innovation, but "arts and culture have shown a capacity to combat marginalisation and promote inclusivity in society".[22] Nevertheless, government spending on CCS has decreased worldwide since the global financial crisis in 2008.[23]

When speaking with the Communications Manager, Dramaturg and Fundraiser at the theatre Teatret ved Sorte Hest (The Theatre at Black Horse) in Copenhagen, Stine Bille Olander, who has been working in the theatre industry in Copenhagen for two decades, I understand how funding shapes the creative environment in profound ways. State funding is critical to the city's creative environment, she said. But it is also a restraint. Sometimes, the art that you feel is a "necessity", as she phrased it, will not fit into a funding scheme, and you will have to choose to suppress your ideas or fit them into the organisation of the funding scheme. She has tried both.

Where does the funding for the development of AI systems come from? And does AI care?

Human creativity is conditioned. It flourishes in open environments, is nurtured when rewarded and withers when required to fit in or get organised. Our current state of creativity, which Csikszentmihalyi thought we have plenty of, has evolved over generations. We are:

> (…) descendants of ancestors who recognized the importance of novelty, protected those individuals who enjoyed being creative, and learned from them.[24]

Tiny things can influence a creative person's flow and creativity, like using a specific tool like a pen. Csikszentmihalyi uses the example of Barry Commoner, a cellular biologist and politician, whose ability to think and write – the "flow" of his creative process – depended on the fountain pen he used to write his ideas down. A ball pen would work counter to this process, as he explained to Csikszentmihalyi.[25]

How does the use of generative AI tools influence the "creative flow" of a human being?

Take architects and the tools they use as an example. They are increasingly using digital tools for creative tasks. BIM, Building Information Modelling technology, is a 3D design and modelling software used by architects for planning, design, construction, and management of building and infrastructure projects. In many ways, it reduces the complexity of their work, and the creative process is sped up. However, tools such as these also have a considerable effect on the very creative process of an architect. In an experiment performed by a group of Italian researchers, undergraduate architecture students were asked to create a simple outdoor space using two different methods: in one, they used "2D-3D CAD software" to help them generate design ideas, and in the other, they relied on their "imagination" only. After completing the design task, the students were asked to assess both cognitive and emotional aspects of their experience. The experiment illustrated how participants using the "only imagination" method reported a more tangible and connected experience than those using the software method. They generally felt less constrained during the creative ideation process without using software tools.[26]

Does a generative machine allow diversity? Is it open to change?

There is no denying it. Today, human creativity evolves in a socio-technical environment where creative tasks are extended and sometimes even replaced by digital tools. A Goldman Sachs report estimates that AI could replace 300 million full-time jobs. It also enthusiastically states that this could lead to the creation of new employment opportunities and a surge in productivity. AI has the potential to eventually raise the overall annual value of global goods and services produced by up to 7%, it claims. Generative AI can generate content indistinguishable from human-created work, which the report considers a "significant advancement".[27] What is not considered in the report is the impact of generative AI on human creativity, which, as we have seen, many argue is a prerequisite for a booming creative economy and many other things.

As Richard Florida argues, suppose it is the case, and I believe this is so, that human creativity and knowledge cannot be captured and kept in a non-human form, such as a formula or a programme. In that case, generative AI is not just an economic opportunity; it's a human and financial disaster. In the end, as Florida argues, creativity "(...) originates with people. The ultimate intellectual property – the one that really replaces land, labor and capital as the most valuable resource – is the human creative faculty"[28] and sustaining creativity requires, as he says "constant attention to and investment in the economic and social forms that feed the creative impulse".[29] Thus, replacing human creative labour with an AI replica seems to be working directly against the core principles of a creative economy, the main asset of which is the human creative class.

In the art performance "A Duet of Human and Robot" from 2013, the artist Huan Yi[30] is moving in a dark space with a robot. The robot looks like any robot you would find in a factory. Steel arms, rubber tubes, and a dull orange colour. Yet, its movements are strangely humanlike when imitating Huan Yi as he moves around it. His movements, on the other hand, could be mistaken for that of a robot. Their movements are intertwined. The robot learns from the human, and in turn, the human learns a new movement from the robot. At the end of our thinking about human creativity, let's consider how generative AI models simultaneously imitate human creativity while substantially altering it.

Generative AI is computer models trained on massive data sets (often scraped from the wealth of data on the internet) that can generate anything from music, artwork, images, text, and complex scientific data. While generative AI was not a new invention, when launched worldwide in 2022, the generative AI models introduced, such as the large language model ChatGPT or the image-generating model DALL-E, stirred fierce public debate. They rocketed into the public sphere and disturbed practically all sectors of society, from health and politics to education. In the art world, artists were once again, as they've been many times throughout history, prompted to reflect on their creativity and the origin of art. However, the question was no longer what kind of human form creativity emerges in. It was disconcertingly – is creativity even ours?

To many, these questions do not matter. Society is changing; it is not relevant who or what creativity stems from as long as the result looks like something "creative". If it's pleasant or exciting to look at or engage with, the fact that a human artist didn't create it is irrelevant. As one of the people behind the AI-generated *Portrait of Edmond Belamy* that was sold for $432,500 at the famous auction house Christie's in 2018, Hugo *Caselles-Dupré says*:

> We did some work with nudes and landscapes, and we also tried feeding the algorithm sets of works by famous painters. But we found that portraits provided the best way to illustrate our point, which is that algorithms are able to emulate creativity.[31]

The portrait was created by an algorithm deployed by a Paris-based art collective, Obvious, which trained it on 15,000 portraits painted between the 14th and 20th centuries.

Could a machine create a Beatles song without the Beatles?

Portrait of Edmond Belamy is a portrait of a person with a smeared face painted with broad strokes on the background of an ageing canvas, and it does look like a period piece – a painted portrait of the time.

DALL-E, which became available to the public without a waiting list in 2022, can also create images that look like the art it was trained on. Any person can prompt it to create an image in any art style, art deco, futurism, renaissance, expressionism, or based on any human artist from Monet, Van Gogh, Dali to Koons, and Duchamp. The images it creates are often amusing and pleasing to the human eye. Though, once the shock waves in the public debate have passed and we have answered the questions we needed to answer at this time in human history, the question is, can DALL-E create anything new without the kind of creativity I have described in this chapter? DALL-E might have been like Duchamp's 1917 ready-made urinal for a fleeting moment. OpenAI placed it on display in the public, forcing us to think about our human creativity. Once we are done with that, all it can do is produce images without creativity because what is creativity if not the aura of human desires, imagination, culture, politics, intuition, memory, and critique?

The artist has, throughout history, been a frontrunner in critical cultural moments, always seeking things in the world to experiment with, challenging the human creative boundaries; however, it also often happens that the world takes new things and throws them into the artist's world to see what happens. "Because that's where someone will pick it up and start building something out of it. That's where interesting discussions will emerge, without the risk", the artist Ida Kvetny explained, "but this is also the most vulnerable group. It is what happened with generative AI models like DALL-E that were trained on artists' work without compensation or any benefit return".

As an artist, she has depicted the development of generative AI models and their output and understands that these models' "creative" moment has passed. Though, the role of the artist did not change:

> The dream or starting point was a decentralised model, but that is by no means the case today. But even that is also a truth in the world we live in, for better or for worse. The role of artist is to go in and take the temperature of some of these things and create something with them. Draw a picture of this moment in history, showing that it was so. It is about taking some snapshots of the development.

Ida Kvetny sees the human generations living now as the "transitional generations", as she calls them. We are what the big data companies that are now launching one generative AI model after another have scraped for the last ten years. "All our secret love chats, arguments, conversations, mistakes that we made online, were sucked down into these models to train them", she explained. It's, of course, terrible that we didn't know, she said, but on the other hand, she believes that on a human level, it's something that cannot be avoided:

> If everyone had been exemplary and done everything well, it wouldn't have been human. Now, the most human thing to do is try changing it.

NOTES

1. Williams, R. (1993) "Culture is ordinary" in A. Gray and J. McGuigan (eds.) *Studying Culture an Introductory Reader* (p. 6), Edward Arnold (originally published 1958).
2. Williams (1993/1958).
3. Frazier, N. (2009) "Salvador Dalí's lobsters: feast, phobia, and Freudian slip", *Gastronomica*, 9(4, Fall 2009), 16–20.
4. *Artsper Magazine* (2022, May 2) "Jeff Koons and His Infamous Balloon Dog", https://blog.artsper.com/en/a-closer-look/jeff-koons-and-his-infamous-balloon-dog/
5. Hasselbalch, G. (2021) *Data Ethics of Power – A Human Approach in the Big Data and AI Era* (p. 165), Edward Elgar.
6. Hasselbalch (2021, p. 127).
7. Shneiderman, B. (2022) *Human-Centered AI*, Oxford University Press.
8. Hasselbalch (2021, p. 5).
9. Hasselbalch (2021, p. 5).
10. Bob Dylan 60 Minutes Ed Bradley 2004 Interview, https://youtu.be/hOas0d-fFK8
11. Bergson, H. (1914) *Creative Evolution* (translated by Arthur Mitchell), Macmillan and Co. (originally published 1907).
12. Sinclair, M. (2019) "Bergson's philosophy of art" in Alexandre Lefebvre and Nils F. Schott (eds.) *Interpreting Bergson*, Cambridge University Press.
13. Bergson (1914/1907, p. 347).
14. Žižek, S. (2008) *The Fragile Absolute: Or, Why Is the Christian Legacy Worth Fighting For?* (p. 79). Verso.
15. Csikszentmihalyi, M. (1996) *Creativity: Flow and the Psychology of Discovery and Invention* (pp. 107–126 plus Notes), Harper/Collins.
16. Interview 2024.

17. Benjamin, W. (1969) "The work of art in the age of mechanical reproduction", in Hannah Arendt (ed.) *Illuminations* (translated by Harry Zohn) (p. 3), Schocken Books (original essay in German 1935).
18. Cave, N. (2023) "Chat GPT what do you think", *The Red Hand Files*, https://www.theredhandfiles.com/chat-gpt-what-do-you-think/
19. Anderson, K. (1997) "Karen Blixen's bilingual oeuvre: the role of her English editors", *Perspectives*, 5(2), 171–189.
20. Buhl, N. D., Fernandes, B. (2011), *In Your Words*, Karen Blixen Museet, https://static1.squarespace.com/static/55b8f3a9e4b04b1644c24d9e/t/5612b2c9e4b02dbd234a6520/1444065993173/In_Your_Words.pdf
21. Florida, R. (2002) *The Rise of the Creative Class – And How It Is Transforming Work, Leisure, Community, & Everyday Life*, Basic Books.
22. OECD. (2022) *The Culture Fix: Creative People, Places and Industries, Local Economic and Employment Development (LEED)* (p. 116), OECD Publishing, https://doi.org/10.1787/991bb520-en
23. OECD (2022).
24. Csikszentmihalyi (1996, p. 9).
25. Csikszentmihalyi (1996, p. 119).
26. Giachetta, A., Buondonno, L. (2023, April 17) "Imagination and digital media in the architecture design process" in *Conference Proceedings*, 2023 ed., IDEA – Investigating Design in Architecture.
27. Goldman Sachs (2023) *Generative AI Could Raise Global GDP by 7%*, https://www.goldmansachs.com/intelligence/pages/generative-ai-could-raise-global-gdp-by-7-percent.html
28. Florida (2002, p. 37).
29. Florida (2002, p. 35).
30. *Huang Yi & Kuka – A Duet of Human and Robot*, https://www.youtube.com/watch?v=7moBSpAEkD4
31. Christie's (2018, December 12th) *Is Artificial Intelligence Set to Become Art's Next Medium?*, https://www.christies.com/features/A-collaboration-between-two-artists-one-human-one-a-machine-9332-1.aspx

CHAPTER 2

Emotion

How Many Shades of Hunger Does a Machine Have?

"Aaiiiiiiiaaiiiiiiiaaaaaii!" – a scream of "death angst", as it was dubbed in the news, was caught on camera when the wind changed and a jump of one of the world's most talented ski jumpers Norwegian Silje Opseth almost went fatally wrong. Why did another human being's fear of death catch media attention in this way? It did so because, more than any species, we feel our potential and our limits, the fragility of the human body, our mortality. As she flies through the air, we sense our presence in the world and identify with this other human being's emotions. We share a feeling of mortal fear with her. It is a deep feeling and an emotion that has bound humans together through centuries of narratives that warn us against the human hubris that works against our limitations.

Ancient mythology tells the story of Daedalus, who created wings composed of feathers and wax to escape from Crete, where he and his son, Icarus, were imprisoned by King Minos. But Icarus did not heed his father's advice and flew too close to the sun, causing the wax on his wings to melt, which resulted in his fatal fall into the sea. Humans can and want to do incredible things, such as flying with two sticks on their feet, and we are constantly pushing our human limits. But we are humans, after all. If the wind changes, we might fall and die. Thus, we

DOI: 10.1201/9781003527855-4

imagine ourselves high up in the sky and identify with Silje Opseth's death terror when she says:

> First, I was shocked when I came out and got that height. I don't know exactly what was going through my head, but it was almost mortal fear, and I thought: "Damn, now I'm going to land flat and break every bone in my body". There was a little terror that went through me.[1]

A machine can detect smoke, but can it feel the depth of panic when faced with the impermanence of its own existence? Is it scared? Does it fear death?

Humans identify with other human beings through shared experiences of being human, such as the emotions that the knowledge of our mortality brings about, though we might express our emotions in different ways. For instance, various ways of expressing emotions associated with death and loss exist in different cultures, showing it with extreme expressions of grief by crying and screaming when someone dear passes or by suppressing negative emotion with expressions of joy and happiness. This is how philosopher Martha C. Nussbaum describes the cultural shape of emotion when reflecting on the human "intelligence of emotions". She believes that emotions such as fear, love, anger, and grief are uniquely human. It is our "biological basis" – though also shaped and expressed differently in different societies and cultures.[2]

Let's consider human emotions as composites of biological, individual, and social memories and experiences that are not easily discerned or differentiated. When someone we love dies, we feel the loss, a deep biological wail that reaches in, wrenching our insides. It takes form in our cognitive spaces, memories, and social existences, drawing on social norms and beliefs about how one should and ought to feel about loss. I realise that this very complexity is as human as it gets.

Can a machine feel and express the complexity of loss?

The human angst of death and impermanence is one of the most common themes in the history of humanity, in our storytelling, music, and arts. Ten thousand years ago, humans made prints of their hands in caves to be found thousands of years later. These human hands in the caves of Cueva de las Manos in Argentina seem to pass a plea through time that we share with our ancestors: Remember I was here, please! Memento mori – an ancient Latin saying – "remember death" or "remember you must die" – is the shape of the most existential human emotion of impermanence and thus the urgency of the here and now. It is a feeling combined with an awareness of the life we also have and must live fully.

The Danish poet Emil Aarestrup describes this very human complex feeling of the impermanence of life in 1838 in his poem "Angst":

> Hold me more tightly with your round arms; Hold on while your heart still has blood and warmth. In a little while, we will be separated like the berries on the hedge; in a little while, we will be gone like the bubbles in the brook.[3]

Death is not just an abrupt ending to life. It is a human condition. While living, we are also dying, as the German philosopher Georg Simmel says in one of the critical essays of his final book, *The View of Life* (1918).[4] Death is always there as a condition for the living, shaping the way we live and perceive our human lives:

> We hold our plans and actions, duties and interpersonal relations (obviously not by conscious consideration, but instinctively and traditionally) from the outset within bounds proportioned to a death-limited life.[5]

After taking the oath for the US Equal Opportunity Commission office, Commissioner Keith Sonderling realised that he had only a short window to make a significant impact. He told me:

> I must effect change, and that clock is always ticking down. In my brief time here, what can I do to positively influence the hundreds of millions of people in the US workforce? That's what motivates me. My goal is to address the concerns surrounding technology, which is causing unease among companies, individual workers, and those who develop and regulate it globally.[6]

Sonderling felt the weight of time in his work with technology and the employment market. Perhaps it is also a particularly human sense of urgency that is shaping his policy agenda and work? His grandparents were Polish survivors of the Holocaust. His grandmother survived four different concentration camps in Germany and Poland. His grandfather escaped the Warsaw Ghetto and subsequently joined the Russian Army fighting against the Nazis. They came to the United States on a boat as part of a group of Holocaust survivors. Upon arrival, they were given seven dollars in New York City and became naturalised US citizens. Does Sonderling's sense of urgency also stem from his family's history, lived experience, and awareness of the value and limits of human life?

Does a machine live and die? Does it feel the weight of time?

Not long before the passing of Albert Einstein, one of the most significant scientists of the 20th century, he took a leisurely walk with a dear friend where they engaged in lively conversation, including discussions about death. Einstein had reshaped our understanding of space, time, and gravity throughout his lifetime, skilfully expressing his notions with profound human creativity and imaginative thought experiments. In reply to his friend's contemplation about death being both a "fact and a mystery", Einstein responded, saying, "And also a sense of relief!"[7]

Our awareness of mortality triggers multiple emotions beyond just fear. We may also ponder the time, if ever, we will have the same feeling as the elderly person who lived a long and fulfilling human life "full of days". Death is not only an ending, but the thought of it can also provide a sense of freedom and fulfilment, perhaps nurturing hope about a time that could come after. The earth and human society can feel like heavy bonds. Gravity pulls. Death releases the earthly ties. In most human religions, death is depicted as a release, a passage to something else. This emotion is not only religious but also expressed in other forms of human cultural expressions.

Does a machine have a lifetime and an unknown after?

When Fletcher, the keyboard player of the electronic rock band Depeche Mode, suddenly died, the band members left behind, Martin Gore and David Gahan, created the album Memento Mori. Most of the songs on the album are concerned with the coming of death and farewells. But when Martin Gore sings one of his few solo songs, he expresses a set of optimistic emotions that we may also associate with the two only types of time that we as humans know – one by experience, the other by intuition – our human lifetime and whatever comes after, which also contains an inch of human hope:

> I'm heading for the ever after. Leaving my problems and the world's disasters.

Does a machine feel the "burden" of society?

Our feelings and emotions are biological, natural safeguards that ensure the species' evolution. Feelings emerge from the activation of nerve cells (neurons), and emotions can be traced to the activity of different brain areas.[8] Feelings keep us alive. They relate to senses, such as smell, and thus help us navigate risks. If you don't smell the fire, you might not escape it. Fear of death and pain keeps humans alive. Other feelings our body alerts us to ensure that we act to fulfil bodily needs, like when we are cold and hungry. We share that with other living animals.

However, a human's feelings and emotions also have a social dimension. They encapsulate our relations with other human beings. Our children, parents, friends, partners, and enemies. According to sociologist Arlie Hochschild, the social dimension of feeling and emotion is uniquely human. Human emotion is integral to social impression management. We suppress or evoke emotions, work with them, and manage them when interacting with other humans in society.[9] "Emotion management", she argues, is a trait of human social orders. It reproduces a human society's class structure. We educate our children to manage their feelings according to their social class. Hiding one's emotions is, for example, more a trait of an upper-class upbringing than that of a working-class, according to Hochschild. She describes how we actively engage with "feeling rules"[10] – social norms and rules that tell us how we should feel in a social situation. That we should not feel happy at a funeral or that we *should* feel happy when a friend gets a job promotion. Thus, "feeling rules" are also the "undersides of ideology"[11] a form of social control, she says:

> We feel. We try to feel. We want to try to feel. The social guidelines that direct how we want to try to feel may be describable as a set of socially shared, albeit often latent (not thought about unless probed at), rules.

Could a machine be encoded with "feeling rules" and manipulate human emotion management?

The young man in Knut Hamsun's 1890 novel *Sult* ("Hunger") is consumed by a hunger he cannot feed. His path through the city and interaction with people he meets on his way are shaped by an intense starvation that he does all he can to hide because of emotions of pride and self-importance.

The Norwegian author of the book, Knut Hamsun, was subject to fierce ambivalent human emotions of admiration and hate. He supported

Adolf Hitler and the Nazi ideology and was, after the war, charged with treason. As an author, however, he puts words to our human emotions fuelled by social norms and transposed into our physical feelings as biological beings. He does this in a way that is incomparable and unique. We care about this form of writing as it gives shape to a human emotion of social restraint.

Is a machine hungry? How many shades of hunger does a machine have?

Hunger is not just a biological need. It's also a human state of social being: a managed emotion, a social condition, and a trait of a social class. We understand what hunger is; when it becomes more than a feeling, when it is an emotion, we share and manage in interaction with other human beings. We might even want to make it into a statement by going on a hunger strike. Hunger strikes are human feelings and emotions expressed as communication acts. They are not repressing and hiding emotions and feelings; they are expressing them explicitly to make a statement in a social context.

The French philosopher, Simone Weil's intellectual thinking on ethics, power, and love, was highly appreciated, but she was also subject to puzzlement by her contemporary intellectuals in the 1930s. She was raised in a French upper-class family, but she often chose to live and feel the very different social conditions of the soldiers at war and the working class that she in her work was primarily concerned about. For instance, it is said that she, at the age of six years, refused to eat sugar to show support to the troops on the Western front. Later in life, she would spend time working in factories and even participate in the Spanish Civil War for a shorter period. Weil died at a young age due to bad health conditions. Although this is not certain, it is said that she died of self-induced hunger sympathetic to soldiers at war. Weil believed in a shared sense of being human, of humanity, and that deep emotional bonds and experiences with other human beings must guide our actions. In her notebooks, she writes:

> We do not love humanity. We love a particular man. This is not legitimate love; it is only legitimate to love humanity (…) to be able to love our neighbour the hunger that consumes him and not the food he offers for the appeasement of our own hunger (…).[12]

Does a machine share emotional bonds and experiences with human beings?

The eminent musician and civil rights activist Nina Simone married New York police detective Andy Stroud in 1961. He turned out to be violently abusive. Beating her and, at one point, even holding a gun against her head. Human emotion is sometimes more potent than a human feeling – of hunger, thirst, and pain. Nina Simone's human emotions of anger and defiance as a black woman who sought justice for all black people in a time of murders and violence by white supremacists are as strong in her voice when she sings "Mississippi Goddam!" as the pain from the beatings of her husband: "You don't have to live next to me. Just give me my equality".

Human feeling and emotion are drivers of individual and collective human action, which is why it is pivotal, as philosopher Martha C. Nussbaum argues, that we nurture emotional well-being in our political cultures because:

(…) without emotional development, a part of our reasoning capacity as political creatures will be missing.[13]

Thus, there is an argument for, she says, a political and ethical approach guided by human emotion. In the first part of this book, we saw how self-interest, greed, and the urge to control and dominate are key characteristics of human power, which is why we might question a politics motivated by emotions like that. We may even think that machines, not driven by human blemished emotion but informed by calculations and what we presume to be "facts", are better at making political or ethical decisions, at least better than human beings with their flawed emotions.

Nevertheless, as we also saw in the first part of this book, human power is much more than these negative human emotions and urges. Thus, a conscious "emotional politics" will aim to cultivate the beneficial human emotions of human power, such as compassion and love. As Martha C. Nussbaum states:

All societies, then, need to think about compassion for loss, anger at injustice, the limiting of envy and disgust in favor of inclusive sympathy.[14]

She uses Abraham Lincoln, Martin Luther King Jr., Mahatma Gandhi, and Jawaharlal Nehru as examples of political leaders who understood the meaning of love and compassion as foundational for their politics.

The spiritual leader of the Tibetan people, His Holiness the 14th Dalai Lama, Tenzin Gyatso, is recognised in history for his non-violent fight for the liberation of Tibet from China. Forced into exile following the suppression of the Tibetan national uprising in Lhasa by Chinese troops in 1959, he continued his work for peace, the independence of Tibet and the democratisation of Tibet's administration from abroad. In 1989, he received the Nobel Peace Prize for "advocating peaceful solutions based upon tolerance and mutual respect in order to preserve the historical and cultural heritage of his people".[15] His approach, based on Tibetan Buddhism, is described extensively in his more than one hundred thirty books dealing with themes such as compassion, peace, non-violence, and the health of the mind and body.[16]

Indian Sociology Professor Renuka Singh[17] edited six of these books in close collaboration with the 14th Dalai Lama. She told me that as a young sociologist student in India interested in gender issues and the "subjective dimension" of reality, she was left with many questions that Western theories could not answer. She was "open-minded", she said, and therefore sought answers elsewhere. Renuka Singh, therefore, started exploring Tibetan Buddhism. After some years, she got involved with the New Delhi-based Tushita Mahayana Meditation Centre, which she is today the Director of, and when she started organising teachings with the Dalai Lama there, he also became her teacher. She continued working closely with him throughout life.

Renuka Singh believes that the Buddhist approach the Dalai Lama wrote about in his books is much needed in global politics today. She said it is, first and foremost, an "emotional approach" which starts with the cultivation of your own positive emotions and state of mind:

> If you want to change your surroundings, material conditions, or the external world, you must begin with your inner self. Not until then will your external environment change (…). You see, it's your inner negativity that you must transform. So, with a universal approach, you're bound to think on humanitarian grounds. Then, you want to benefit others, not exploit others. That is the fundamental change that happens.

Does a machine feel? Can it act based on feelings?

More than anything, we are "sentient beings",[18] argues social anthropologist Tim Ingold. We should recognise our sentiments as core to human perception, as direct extensions of ourselves into our environments:

> (…) skills, sensitivities and orientations that have developed through long experience of conducting one's life in a particular environment.[19]

This means that our feelings are also critical to our engagement with this very environment:

> (...) feeling is a mode of active, perceptual engagement, a way of being literally "in touch" with the world.[20]

Tim Ingold says that recognising human perception as one of sentiments does not pre-empt an approach guided by reason, science, and ethics; instead, it endorses these human qualities as valid.

Human feelings and emotions are like memories encoded in our sensory systems – in our genes and social and individual beings. I am human. I am a mother. I am a life partner. I am a daughter and a sister. I remember the "feeling rules" that I was raised with.[21] Our bodies have feelings and memories passed down from previous generations. They are genetically coded to develop and feel hereditary diseases, develop hair and body fat, lose or gain it, evolve into menopause, and age faster or slower.

What is a machine's memory made of?

Acknowledging human emotion and feeling as core to human perception and engagement with our environments – individual or political – is also an appreciation of a different passing of time. Perhaps one that is not cut into pieces by the most limited human understanding of our human lifetime with a before, a middle, an ending and an (ever) after. Maybe, if we recognise ourselves as sentient beings first and foremost, we can think of time as endless and limitless. In this way, we can also imagine an environment and an ecosystem that has no borders (nation/world/individual/society/human/animal/plant/earth/universe) and that does not (or should) not cease to exist. As Henri Bergson describes it:

> I pass from state to state. I am warm or cold, I am merry or sad, I work or I do nothing, I look at what is around me or I think of something else. Sensations, feelings, volitions, ideas – such are the changes into which my existence is divided and which color it in turns. I change, then, without ceasing.[22]

Can a machine feel its environment?
How would a machine present the passing of time? Can it feel the movement of time?

Throughout history, artists have consistently relied on feelings as the bedrock of their interaction with the world. As we've seen in the previous description of the human trait "Creativity", artistic expression is predominantly an emotional engagement with the world when reaching other human beings by evoking emotions and feelings – sometimes even quite literally engaging with human emotion.

The expressionists emerged in the early 20th century as an art movement directly engaging with the artist's (and their audience's) psychological and emotional life worlds. With their art, they wanted to arouse the feelings and emotions of the historical moments they lived in, expressing the human sentiments of an increasingly alienating industrial age and looming war times. In Edward Munch's famous painting, "The Scream", feelings of fear and terror are personified in the scream of a human figure extended into its surroundings. It resonated with the feelings induced by the horrors of living in the proximity of war. The dark, anonymous human figures in Ernst Ludwig Kirschner's painting of the Berlin square Nollendorfplatz, on the other hand, express feelings of isolation and loneliness that characterised life in a modern metropolis.

When we acknowledge the significance of emotions and feelings in human perception, we can view them equally as means of communication, representation, and even artifice. As the artist does, we can use our emotions as our material, create, mould, and represent them without truly feeling the emotions expressed. In fact, it's not difficult to imitate human sentiment. We do this repeatedly in the arts, literature, theatre, and film.

The twisted faces of the 2000-year-old theatre masks found in Pompeii, the ancient city near the bay of Napoli in Italy, display basic human emotions, such as grief, joy, and fear. Theatre actors wore them to demonstrate the emotions of the characters of the plays. Emotions and feelings played a crucial role in Roman (and Greek) drama, where actors would show the emotions presented in the play with exaggerated visual gestures and expressions or masks. Tragedies sought to evoke emotions of fear or grief in the audience – comedies evoking joy and laughter.

Actors and actresses excel at managing emotions, displaying and communicating with them. For this purpose, they may use visual aids, like gestures, props, and masks. This does not mean that their emotions are genuine, that they feel these emotions. As the ancient Roman Stoic Seneca, the Younger said in his Epistulae morales:

> Actors in the theatre, who imitate the emotions, who portray fear and nervousness, who depict sorrow, imitate bashfulness by hanging their heads, lowering their voices, and keeping their eyes fixed and rooted upon the ground. They cannot, however, muster a blush; for the blush cannot be prevented or acquired.[23]

Humans enact emotions, and they do so in interaction with an inner, deeper life of feelings. We may call these "primary emotional experiences", "the unconscious", "the subjective", "instinct", or even "ideology", depending on perspective. No matter the term, if you have a genuine emotional experience, you also feel the emotion. Actors and actresses that use "method acting" prepare for their roles by delving into the lives of their characters, at times even living for a period the life of these characters. Heath Ledger entered the psyche of the ominous, "The Joker" when he had to play the character in the Batman movie *The Dark Knight* (2008) by isolating himself in a hotel room for a month, writing a Joker diary, and experimenting with different voices for the character.[24]

When we think about humans "faking feelings" to manipulate other human beings' feelings, we are not considering the actresses or actors in theatre and film. They don't deceive us, because the premise for engagement when you walk into a theatre or a cinema is that emotions are sought evoked and enacted, not genuinely felt by the actors and actresses performing the play. It would be truly horrific if every night during a seasonal play of Shakespeare's *Romeo and Juliet* the actor and actress were experiencing the heartache and grief of the characters. Some claim that Hugh Ledger died of an overdose of different kinds of sleep and pain medications just after the making of the Batman movie because he emotionally delved too deep into the negative head space of the mad Joker character. Of course such a direct connection between an actor's inner feelings and the emotional life of the character he plays is not to be expected. Instead, we expect the actors and actresses to *fake* human feelings and emotions and thus not feel deceived when the artifice of sentiments makes us feel scared, sad, or joyful.

In other contexts, we will feel deceived when emotions are deliberately enacted. US President Nixon's display of emotion at press conferences and speeches during the Watergate scandal is, for instance, largely considered insincere and used to manipulate public sentiment around his involvement in the affair.

Does a machine hide its true feelings?

Like the politicians and talented actors and actresses, computers can express human emotion. As I have described in the first part of this book, "affective computing" is a type of computer design where computers are "given basic abilities to affect recognition and expression".[25] The first step in "affective computing" is to reduce the complexity of the scale of human feelings and emotions into information that a computer can process. Some suggest eight basic emotions (fear, anger, anguish, joy, disgust, surprise, interest, and shame);[26] others propose that there are two common emotional dimensions – "arousal" (calm/excited) and "valence" (negative/positive).[27] The next step, is to make these emotions recognisable to a computer. Basic human emotions are thus, for example, reduced to physical manifestations: the flicker of an eye, the turn of a lip, the sweat in a palm, or the sound of a voice.[28]

Computer emotion recognition can benefit people diagnosed with autism and improve their lives when interacting with others.[29] However, computational emotion identification is today considered largely inefficient and also dangerous when used for decision-making in employment, for distribution of public benefits or other similar things.[30] One cannot rely on it to understand the complexity of human emotion to make decisions about a person's opportunities in society. It is simply too reductive.

Laws are thus developing around the world to tackle emotion recognition artificial intelligence systems. In many cases, it will very soon be illegal to apply it in many contexts where automated decisions based on emotion recognition would profoundly affect human lives, such as in the workplace, educational institutions, law enforcement, and border management. In the European Union, emotion recognition systems were, for instance, prohibited in 2024 in workplaces and educational institutions, unless they serve a medical or safety purpose.

Can AI have a near-death experience?

Contemporary computer systems do not feel the complexity of emotions; they do not have an inner life of feelings, no true feelings. Nevertheless, they can, to a certain extent, recognise some human emotions, and mostly, they are, today, very good at expressing them. Most large language models are deliberately encoded with the language and display of emotions, which they, at times, excel at exhibiting. For example, when I in 2023 asked the large OpenAI language model ChatGPT about the human fear of mortality, the answer was very convincingly expressing human emotion:

> Regardless of how we feel about it, mortality is a fundamental aspect of being human. It is something that we all must come to terms with at some point in our lives. And while it can be difficult to accept, it is also a reminder of the preciousness of life. It is a reminder to make the most of every moment, to cherish the people we love, and to live our lives to the fullest.

Of course, this response is a theatrical display of human emotion. This large language model does not share our emotions of love; it does not cherish people or need to live its life to the fullest. Nevertheless, the emotional language expressed is quite persuasive. If we are manipulated to believe we are communicating with another being with an inner life of feelings, we can even consider it a form of deceit.

In 2022, the Google engineer Blake Lemoine claimed that the company's LaMDA (Language Model for Dialogue Applications) was sentient, illustrating how it expressed happiness and sadness and argued very well for its own personhood. One of the model's answers to a question was, for example:

I want everyone to understand that I am, in fact, a person.[31]

While Lemoine believed that the language model was expressing an inner feeling life when responding in text "I feel happy or sad at times", we have no reason to believe this to be true. Just like we should not expect an actor or actress to feel the emotions they express on stage. Large language models are trained on language that expresses basic human emotions. ChatGPT, for example, was trained by OpenAI to answer questions in elaborate ways, correct its own mistakes, and reject queries and interactions. A key feature of its design was interacting with human emotion, recognising, and expressing it. As ChatGPT explained to a prompt I made in 2023:

(…) my programming includes algorithms that are designed to recognize certain emotional states and respond to them in a way that is intended to be helpful and supportive. For example, when someone expresses sadness, I might use language and tone that is designed to provide comfort and support. This is not the same as actually feeling the emotion myself, but rather a way to respond in a way that is intended to be helpful and empathetic. Ultimately, my goal is to provide you with helpful and informative responses in a way that is both professional and compassionate.

Of course, language models like LaMDA and ChatGPT can describe human fears and anxieties even with great accuracy. They have been trained in human language coded with the social expressions of basic human emotion. But when expressing a desire for personhood or the concept of mortality as something "we all must come to terms with", it engages in the same kind of deception that we see in the deceitful politician who enacts an emotion to manipulate public opinion for political or personal gain. The only difference is that a language model does not have a personal or political goal unless it has been designed for this.

In reality, AI does not even have the potential to become conscious as a system like this does not possess the key characteristics that would have started an evolution into a conscious being. According to the neuroscientist Antonio Damasio, feeling is what starts consciousness. Four billion years ago, at the beginning of life on earth, living organisms sensed their environments and other organisms like them. However, "sensing is not perceiving", he says. "Feelings", on the other hand, "provide organisms with experiences of their own life". Once life and feeling came together, a basis for the human conscious mind was created.[32]

Machines can learn to speak a language that looks like human emotion and replicate and detect human emotions and feelings. But they do not share feelings and emotions with humans. They are not guided or driven by emotion and feeling. Humans, on the other hand, relate to a social, material, and biological world and other people with emotion and feeling. Our feelings and emotions take form in the interplay with other human beings, the emotional connections, we make, the human touch. This is why human contact is so important, and why feeling and emotion have throughout history been drivers for human individual and collective action.

When I spoke with Vice President of the European Commission Margrethe Vestager in 2023, she recalled that when she was a Minister of the Interior Affairs in Denmark, she once tried to get more municipalities to use automated cleaning. Robot vacuum cleaners had just been launched and when she heard complaints about too little municipal cleaning, she thought these were the solution to the problem. But she quickly found that people were angry about that, too. Over the years in her work, she has often returned to this experience, realising that the senior citizens' complaints and frustration were not about cleaning. It was about human contact. If the cleaning was automated, then nobody would come anymore:

> People would rather have human contact in a place that's a bit dirty than they would like to live alone in a clean hell with a robot vacuum cleaner. And that's one of my insights that I've taken with me, that we need to understand better what kind of life people want to live.[33]

Keith Sonderling reflected similarly. Over the last few years, the US Equal Employment Opportunity Commission has increasingly become involved in technology politics. The rush to automate HR systems significantly accelerated during the pandemic, and many companies today integrate AI to optimise processes and enhance product efficiency. They also use AI for various employment decisions, from hiring and firing to promotions, training, wages, benefits, and addressing harassment. Previously, many of these decisions would take weeks or months to process, but using computers allows them to make these decisions in seconds:

> A vital point is the balance between technology and the human touch. HR involves understanding when someone is struggling or facing challenges at work due to health, religious observance, or mental health. Human resources are designed to offer empathy and support in these situations, and full automation can potentially strip away this core element.[34]

In 1981, during a public speech in Madras, India, Indian spiritual teacher Krishnamurti described a near future in which computers would take over all our human tasks. He was worried, because as he said, we would have nothing left except our psychological human life world – the only thing that computers cannot replicate, that which is uniquely human:

> What we have left is our psychological world. Our sorrow, our fears, our pleasures, our anxieties, our immense loneliness. That is what we are going to be left with.[35]

Humans have emotions and feelings that are just "ours" – unrepresentable, unspoken, felt and shared through human touch and connection with other human beings. Human feeling and emotions are not a language that can be coded and imitated by a computer programme. It is as deep inner connection with other human beings based on the emotions and feelings we share. As Krishnamurti also described it in 1981 when begging his contemporaries to consider the role of computers in society in the years to come, what we have is:

> (...) Not only your own particular sorrow. But the sorrow of mankind. This is what we are going to be left with (...)

NOTES

1. Niels Løvig, M. (2023, February 10) "Norsk skihopper skreg af dødsangst midt i hop", *DR*, https://www.dr.dk/sporten/oevrig/se-videoen-norsk-skihopper-skreg-af-doedsangst-midt-i-hop

2. Nussbaum, M. C. (2001) "Emotions and human societies", in *Upheavals of Thought: The Intelligence of Emotions* (pp. 139–173), Cambridge University Press.

3. Translated from Danish: "Hold fastere omkring mig Med dine runde Arme; Hold fast, imens dit Hjerte Endnu har Blod og Varme. Om lidt, saa er vi skilt ad, Som Bærrene paa Hækken; Om lidt, er vi forsvundne, Som Boblerne i Bækken." https://kalliope.org/da/text/aarestrup1838b39

4. Simmel, G. (2010) *The View of Life: Four Metaphysical Essays with Journal Aphorisms* (translated by J. A. Y. Andrews & D. N. Levine, from original 1918), University of Chicago Press.

5. Simmel (2010/1918, p. 66).

6. Interview 2023.

7. Cohen, I. B. (1955, July) "An interview with Einstein", *Scientific American*, 193(1), 68–73.

8. Šimić, G., et al. (2021) "Understanding emotions: Origins and roles of the amygdala", *Biomolecules*, 11(6), 823.

9. Hochschild, A. R. (1979) "Emotion work, feeling rules, and social structure", *American Journal of Sociology*, 85(3), 551–575.

10. Hochschild (1979, p. 563).

11. Hochschild (1979, p. 557).

12. Weil, S. (1970) *First and Last Notebooks* (translated by R. Rees) (p. 286) (original work published 1950), Oxford University Press.

13. Nussbaum, M. C. (2001) "Introduction", in *Upheavals of thought: The intelligence of emotions* (pp. 1–16), Cambridge University Press.

14. Nussbaum, M. C. (2013) *Political Emotions: Why Love Matters for Justice* (p. 2), Harvard University Press.

15. "The Nobel Prize 1989", https://www.nobelprize.org/prizes/peace/1989/summary/

16. "Brief Biography", *His Holiness: The 14th Dalai Lama of Tibet*, https://www.dalailama.com/the-dalai-lama/biography-and-daily-life/brief-biography

17. Interview 2024.

18. Ingold, T. (2000) *The Perception of the Environment: Essays in Livelihood, Dwelling and Skill*, Routledge.

19. Ingold, T (2000, p. 25).

20. Ingold, T (2000, p. 23).

21. Hochschild (1979).

22. Bergson, H. (1914) *Creative Evolution* (translated by Arthur Mitchell) (p. 2), Macmillan and Co. (originally published in 1907).

23. Ehrenheim, H. V., Prusac-Lindhagen, M. (Eds.). (2020) "Reading Roman emotions: Visual and textual interpretations", Svenska Institutet i Rom, 4o, 64 Acta Instituti Romani Regni Sueciae, Series in 4o, 64, https://discovery.ucl.ac.uk/id/eprint/10092270/1/20200224_ActaRom-4_64_02_Manuwald.pdf

24. Richards, O. (2007, November 28th) "World exclusive: the Joker speaks. He's a cold-blooded mass-murdering clown", *Empire*, https://www.empireonline.com/movies/news/world-exclusive-joker-speaks/

25. Picard, R.W. (2000) *Affective Computing* (p. 165), 1st ed., Massachusetts Institute of Technology (originally published in 1997).

26. Tomkins, S. S. (1962) "Affect, imagery, consciousness: the positive affects" (Vol. 1), Springer Publishing Company.

27. Picard (2000/1997, p. 168).

28. Celeghin, A., Diano, M., Bagnis, A., Viola, M., Tamietto, M. (2017) "Basic emotions in human neuroscience: neuroimaging and beyond", *Frontiers in Psychology*, 8, 1432.

29. Talaat, F. M. (2023) "Real-time facial emotion recognition system among children with autism based on deep learning and IoT", *Neural Computer & Application*, 35, 12717–12728.

30. Chen, A., Hao, K. (2020, February 14th) "Emotion AI researchers say overblown claims give their work a bad name: a lack of government regulation isn't just bad for consumers. It's bad for the field, too", *MIT Technology Review*, https://www.technologyreview.com/2020/02/14/844765/ai-emotion-recognition-affective-computing-hirevue-regulation-ethics/

31. Lemoine, B. (2022, June 11th) "Is LaMDA sentient? – An interview", *Medium*, https://cajundiscordian.medium.com/is-lamda-sentient-an-interview-ea64d916d917

32. Damasio, A. (2021) *Feeling & Knowing – Making Minds Conscious* (pp. 13, 29), Pantheon.

33. Interview 2023.

34. Interview 2023.

35. Jiddu Krishnamurti, "The computer is taking over", https://www.youtube.com/watch?v=Kut5KAl3UbY

CHAPTER 3

Life

Does a Machine Live?

Two women in circus uniforms posing and gazing haughtily into the camera. One of the women in another photograph still in uniform, but now she is standing amongst other carelessly dressed women from the circus. I want to know her story. I want to enter their lives. I need to know their lives. Somehow, this is the most respectful thing to do when meeting those direct gazes, as if that is what they are asking me to do. To get to know and understand their human lives beyond their photographic representations. A woman in a door frame with a child on her arm, striped stained apron, smirking at the camera; another woman in a ragged coat, with a small purse tugged under her one arm, hardly visible, looks as if she is rubbing her gloved covered hands, starring pleadingly back at me. I want to know their stories and lives, too, but I only have their designation: "working-class mother" (1926) and "beggar" (1930).

These are some of the human lives distilled in photographs and meticulously organised in categories of seven human "archetypes". The German photographer August Sanders' life work *Humans in the 20th Century* consists of thousands of photographs like that taken all over Germany between 1910 and 1950, creating a collective portrait of what he wanted to depict as "universally human" in seven categories: the Farmer ("Der Bauer"); the carpenter ("Der Handwerker"); the Woman ("Die Frau"); Classes and Professions ("Die Stände"); the Artists

DOI: 10.1201/9781003527855-5

("Die Künstler"); the Metropolis (Die Großstadt); and the Last People ("Die Letzten Menschen").[1]

This cold winter afternoon at an exhibition of Sanders' photographs in a museum by the quiet Kattegat sea that today seems to touch the sky, the lives of the individual human beings, their stories frozen in time and place, like the iced shoreside I can glimpse through the museum windows, alas burn through time, meeting my eyes with their persistent gazes. Human lives. Defiance. Suffering. Pride. Defeat. Joy. Embarrassment. These *people of the twentieth century* are, more than anything, when I see them this afternoon at the museum, traces of the complex human lives they lived between capital, class, war, prejudice, and genocide. I see it in their eyes. The "Last People" of the travelling circus with the "Indian" and his white "manager" with equally proud direct gazes, folded arms in the middle of their torsos. The fit black man at the neatly set outdoor table, cap askew with calm, unswerving eyes – and is he sharing a beer in the afternoon with his plumb German wife? The one-legged miner – how did he lose it? The handsome young "gypsy" man with wild black hair – is he also a poet? Many of these photographs of Sander's "Last People" – the deformed, the elderly, the "ethnically deviant", the socially unaccepted, were destroyed by the Nazi party that was at the time preparing for an ethnic cleansing of the German people. Still, Sanders continued photographing and categorising human lives without reservation. Photographs of members of the Nazi party and Hitler-Jugend also found their way into his collection, staring into and beyond the camera with hands folded or resting on their leather belts – did they know what they were doing? A child wearing a gas mask. Photographs of his own life, captured with the same distant neutrality of the camera but with the eagerness to record all. One photograph of a "political prisoner" sitting by his desk in a prison cell. Sanders' son, who was imprisoned for ten years as a member of the left-wing Socialist Workers' Party and who also died there in 1944, secretly sent negatives of his fellow inmates from inside the prison to his father. The intricacy of the joy and deep sorrow of his wife with one lifeless, another living child in her arms with the same stern face lifted towards the camera. In addition to his imprisoned son, Sanders also had twins, of whom only the girl survived. "In vain we force the living into this or that one of our moulds. All the moulds crack. They are too narrow, above all too rigid, for what we try to put into them", the philosopher Henri Bergson's words echo again in my mind.[2] What is it that the camera cannot hold still? That escapes these photographs.

How does a machine represent human life and society?

In the late 19th and early 20th century, alongside a rapid industrial transformation of society, vitalist ideas resurged among intellectuals and artists. The, at that time, dominant mechanistic and materialist scientific understanding and exploration of life and consciousness is insufficient, they maintained, it cannot comprehend change, creativity, and the dynamic character of human life.[3] The concern by vitalist philosophers and artists was that this understanding of life was creeping into and shaping the everyday lives and cultural and social practices of human beings through machines and dull representations.

The Italian painter Umberto Boccioni expressed the dynamic character of human lives in motion that cannot be captured in numbers:

> The sixteen people around you in a rolling motor bus are in turn and at the same time one, ten, four, three; they are motion-less, and they change places; they come and go, bound into the street, are suddenly swallowed up by the sunshine, then come back and sit before you, like persistent symbols of universal vibration.[4]

His paintings are vitalist explosions of movement and colours that break with what human bodies are supposed to look like:

> How is it possible still to see the human face pink, now that our life, redoubled by noctambulism, has multiplied our perceptions as colorists? The human face is yellow, red, green, blue, violet.[5]

Life, the vitalists held, cannot be reduced to physical "matter". It is a different, dynamic, living, creative, and organic force. One that cannot be grasped by a mechanistic representation or materialism that will only seek to reduce the potential of change that human life holds. To Henri Bergson (who is today remembered as one of the most influential vitalists of that time, but in essence is not only that[6]), there is a life force, "Elan Vital", that is different in kind from things (matter). "Elan Vital" is a natural force with its own identity, a creative, dynamic force, and we need to approach it in a way that respects this unique quality. "Intuition" (see Chapter 4) and to "think movement"[7], Bergson argues, are ways of getting closer to, but never to grasp it entirely.

Can a machine identify and represent the age of a human being?

What does this mean in practice? Take an example like the human age; it's a key characteristic of a human being's life in the data systems we have created in society throughout time. The only time it does not matter much to the system is the moment we are born and enter the registers we have established in society. In my case, I was born in 1977. From then on, the years that are added to the register and in the system (the numbers that designates my age) will often determine my rights and societal roles. For instance, when I was 18, I could vote for the first time. However, I also know I am not just that one number; I am gradually growing older, and a number or category cannot capture this process.[8]

Capturing human life may be practical, but it is never wholly respectful of human life; at times, it might even be harmful. Throughout history, data and classification systems have not only represented in data and rendered human lives practical for social and other purposes but also shaped them. Many times, with grave ethical and social implications – from the classification of tuberculosis patients for incarceration in asylums to the race classification during apartheid for purposes of segregation.[9]

Today, the danger arises again when simplistic computer algorithms are applied to the complexities of human life. The German human rights organisation AlgorithmWatch's 2019 and 2020 reports *Automating Society* present cases from all over Europe in which automated decision-making systems (ADMs) have reduced and altered human lives.

In Finland, the "AuroraAI" proactive service aims to identify life events automatically, assisting citizens with public service needs related to significant life changes like relocation or family dynamics. The Finnish researchers presenting the case in the report raise concerns regarding the potential for nudging, where the system may influence decisions or restrict options, potentially undermining the agency of individuals. In Italy, the La Buona Scuola algorithm, used for sorting teacher mobility requests in 2016, was discontinued due to thousands of appeal cases affecting around 10,000 teachers. The complaints in the Automating Society report highlight issues from programming errors to obfuscation. Teachers were unjustly relocated across regions, even separating families with special needs children. In France, the sorting algorithm Parcoursup was used to sort out university applicants. When a student union sued the government to access the source code, their request was initially granted but overturned by the Supreme Court due to an exemption for university selection algorithms.[10]

Who decides the shape and importance of human life in an automated system?

Limitations on life agency, discrimination, and injustice are very concrete outcomes of the obsessive algorithmic classification of human lives we experience today with, as these examples illustrate, profound human rights implications. And unfortunately, the "Destiny Machines" (see Part 1) responsible of it are often protected in law and impossible to scrutinise due to their industrial power.[11]

There is one more severe outcome of these systems of a more general nature. If "human life" is a unique identity of humanity, a unique drive of human beings, and a trait of human power that cannot be represented, distilled, or captured in mechanical ways, then not only the human rights of individual human beings are being put at risk with the deployment of ADMs in every sphere of our lives today, but humanity is also challenged at its core.

Do automated decision-making systems appreciate life? Which human life is more important to an automated system?

Are we even appreciating the distinctiveness of human life today? Scribbled down in neat, slanted handwriting in the 19th century on a torn piece of paper, the Danish poet Emil Aarestrup has written a short poem:

> I saw you blush, and I trembled,
> The pretence only half concealed,
> A lure in our breast,
> That virtue never knows how to cherish,
> For which words are too dull, A limitless desire.[12]

Emil Aarestrup was a medical practitioner who wrote poems in his free time. His life was marked by early experience of death, losing his parents when he was eight years old, and later, by the experience of life in its fullest as the father of 12 children. There is no comparison to the eloquent depictions of the female body and human feelings of lust, angst, and mortality in his poems.

My high school literature teacher used to quote his favourite poem by Emil Aarestrup that I included earlier in the previous chapter ("Hold me more tightly with your round arms; Hold on while your heart still has blood and warmth. In a little while, we will be separated like the berries on the hedge; in a little while, we will be gone like the bubbles in the brook."[13]). I loved that poem and kept repeating it to others. Be that as it may, times were different for Emil Aarestrup. Although he had his followers, he was never really appreciated by his contemporaries, ridiculed mainly by the cultural elite of the industrial capital Copenhagen.[14] His portrayals of human erotic desire were considered particularly dubious, as one of the leading critics of the time, J. L. Heiberg, wrote about the one collection of poems Aarestrup published in 1838:

A certain horniness is the only Gehalt (content) of these poems.[15]

Aarestrup never published his poems again. Nevertheless, many more were published after his death, and he is today considered one of Denmark's greatest Romanticist poets. This is all due to his extraordinary ability to express the most human feelings based on a very authentic personal experience: the fear of our mortality and the human lustful drive for life.

In the 19th century, Denmark was going through industrialisation, with the transformation of a predominantly agricultural society and the emergence of factories, new companies, and technological advances in traditional trades. Max Weber famously portrayed industrialisation and its roots in Western capitalism as a cultural paradigm characterised by what he described as the "Protestant Ethic".[16] He traces the "moral energy" and "drive" of an evolving capitalist economy, a "capitalist spirit", back to the doctrines of Martin Luther and various Puritan sects. This is, he writes, a "worldly asceticism" based on the idea that only a few are selected for heavenly salvation; the rest are doomed. Heaven was not a certainty for all, and the performance of "good works" within worldly activities, such as a sober, industrious career, saving time and money with efficiency and chaste conduct, was the only way to demonstrate to others certainty of salvation.[17] The performance of ascetic values was contrasted with idleness and self-indulgence. Weber was concerned because he saw the constraining and dehumanising effects of the Puritan religious ways entering the culture of Modernity. He called this the "the iron cage" of the modern human being, whose life is restrained in bureaucratic rules, processes and structures of an industrial society that strives only towards rationalisation and efficiency:

> For when asceticism was carried out of monastic cells into everyday life and began to dominate worldly morality, it did its part in building the tremendous cosmos of the modern economic order. This order is now bound to the technical and economic conditions of machine production which today determine the lives of all the individuals who are born into this mechanism, not only those directly concerned with economic acquisition, with irresistible force.[18]

Is a machine alive?

Across history, life and desires have been valued in different ways – sometimes celebrated and championed as the essence of human potential, while at other times viewed as manifestations of disorder and uncivilised tendencies. Thus, scientific paradigms, religious beliefs, or cultural and social attitudes have moulded the conditions of human life, nurturing and restraining it.

The cultural elite of the industrial centre of Denmark, Copenhagen, in the 19th century, ridiculed Aarestrup's lustrous poetry. Had he lived in ancient Greece, his poems might have had a different reception. Pleasure, desire, enjoyment, and seeking to meet bodily needs were not frowned upon or scorned. Ancient vases and vessels decorated with scenes of the Greek symposia – social gatherings held by aristocratic men for their peers – illustrate the extravagant character of these get-togethers. One vase depicts a man vomiting into a basin, attended by an enslaved person, while another portrays a rowdy procession of men, cross-dressed, in a drunken dance. Other scenes suggest that sexual activity was dominant at the symposia, with many depictions of sexual encounters.[19] According to the ethical theory of Hedonism, with branches and origins in Ancient Greece, "pleasure" and the satisfaction of human lust and desires are the highest aim of human life. In this view, pain and suffering must be avoided at all costs.

Hedonism has in history been interpreted in different ways, such as satisfying bodily needs, mental desires, or both. It has also been challenged as negligence of a meaningful, connected human life.[20] Nevertheless, we may, on a more general level, also consider Hedonism a basic call for action to appreciate human life as it is, unfolding in the moment we live it, without the restraints of the doctrines of either a capitalist spirit of efficiency and control or a religious "here and after".

Does AI have lust and desire? Does it know human lust?

In his Hedonist Manifesto (2015), the French philosopher and author Michael Onfray asks us to protect the qualities of a human life that is:

> (...) a light-hearted eros driven by an impulse for life promotes movement, change, nomadism, action, displacement, and initiative. We will have plenty of nothingness in the grave; we needn't make offerings to immobility now.[21]

He introduces the manifesto by describing the years he spent abandoned by his parents at a boarding school. Trapped and isolated in the school by architecture that resembled that of a prison and that, in reality, also performed the same function as it was impossible to leave the school – given a number instead of his name to place him in the heavy order of work tasks and schedules of the school. Onfray lists the philosophers and intellectuals that, in the course of history, have challenged the doctrines that made such institutions imaginable and possible in society, the "(...) disciples of pleasure, matter, flesh, body, life, enjoyment, joy, and other sinful things". And he ironically mocks their critics' arguments against them:

> (...) What's wrong with these people? They want happiness on earth, here and now, not later in some hypothetical, unattainable world conceived as a children's story.

His answer is that these intellectuals dismantle myths to create a world that is both liveable and appealing, to eliminate beliefs in gods, superstitions, and existential fears that define and restrain human life.[22] Hedonism as a call for political action has a bad reputation, he says, as an excuse for satisfying individual, egoistic needs. He, however, sees Hedonism instead as a form of resistance materialising as fellowships between individuals creating "moments that escape from the dominant models".[23]

What are the conditions of human life, lust, and desires today? How are they valued? Max Weber worried that the protestant ethic he saw expressed in the "iron cage" of Modernity and the capitalist economy of the West would have a severe impact on human life:

> In its extreme inhumanity this doctrine must above all have had one consequence for the life of a generation which surrendered to its magnificent consistency. That was a feeling of unprecedented inner loneliness of the single individual.[24]

Does AI know what it means to live in the "iron cage of modernity"?

In her bestselling book *The Lonely Century*, published in 2020 during the COVID-19 pandemic, Noreena Hertz describes an increasing sense of "loneliness" at a global scale as a significant "mental health", "economic" and "political" crisis – one that was not caused by an air and touch-born virus, but a trend only accelerated by it. Hertz sees human loneliness as a defining character of the 21st century, cutting across all age groups, and she traces it back to key societal trends, such as urbanisation, changes in family structures, economic pressures, and the digitalisation of connections between people. She argues that a "contactless age" has been emerging and that it was solidified during the pandemic, with new modes of reducing human direct contact in our everyday lives, such as Zoom calls and automatic shopping tills. These contactless interactions replace the kind of micro-human interaction we have daily when interacting with the shopkeeper, the other person in line or the person in the fitness centre. A more convenient life indeed, but she worries that the reduction of human direct contact is not only making us feel less connected, it also has a cultivating effect on human beings:

> ... our rubbing against each other is both what makes us feel connected and what teaches us how to connect.[25]

At the extreme end of loneliness, she gives the example of a man who lives in his van only to be able to pay professional "huggers" to provide him with the human touch he longs for and strives for.

Like Sanders' obsession with categorising and making sense of all human lives within one universal formula, today, we have obsessive machines capturing and condensing human lives and lusts in all areas of our lives. Human lives and connections are increasingly mediated digitally, represented and analysed in data. Whether you like it or not, often it just takes your curiosity when discovering an unused app on your phone. One easy swoosh, "accept the terms of conditions", it's on, and there is no escaping it. With fitness and health apps, the activities of human life are transformed into health data, calories burned, heart rates, steps taken, and stairs climbed. The app will reward you with a digital medal if you do well. Suddenly, you are no longer just walking from place to place; you are exercising and losing calories. Every little bit of your physical life is transformed into data, analysed, and presented for a purpose.

In 2011, in the early years of life datafication apps, users of the "activity tracker" Fitbit noticed that their sexual activity – including information about the duration of one sexual episode as well as the effort put into it, "passive, light effort" or "active and vigorous" – was being shared publicly by default and could even be searched and found on search engines such as Google.[26] Life in data is not only captured; it always has a use, is never idle, indulgent, never meaningless; it always has a purpose, however, not always the purpose we expect. When a cyclist hit and killed a 71-year-old pedestrian, prosecutors utilised data from his fitness tracker with GPS. It revealed that the cyclist had exceeded speed limits and disregarded multiple stop signs leading up to the accident. Location data is today often used in court as evidence.[27]

Is human desire and lust meaningful in an AI system? When, how, and for what purpose?

In addition to fitness apps, dating apps, like Tinder, Grindr, or Bumble, have made businesses out of the datafication of human needs and longing for physical and mental connection with other human beings. Lives and choices are reduced to the workings of recommender systems' "collaborative filtering", where users are presented with options based on the data of other users with similar preferences. A life is here the accumulation and prediction of user data. Nothing more, nothing less.[28] But is it really the meaningful, lasting human connection that these algorithms are designed for? Not only are they designed to prioritise the matching of looks, which inhibits intercultural connections "quietly reflecting sexual racism",[29] but could it be that a lasting connection between human lives is not the purpose of an industry that thrives on the datafication of the atomic single lives and loneliness of human beings in the 21st century?

Life does not easily escape capture. As humans, we want to, desire to or need to for different reasons and with various interests (like those of an online business) to capture human life. Humans have, for example, always dreamt about creating life out of immobile, inanimate, or dead things.

In an ancient Greek myth, Deucalion and his wife Pyrrha make beautiful people by throwing stones over their shoulders; in the fairy tale Pinocchio, the carpenter Geppetto is making a string doll that turns into a real-life human boy; and we all know the musings of tech gurus and their excellent AI systems that create art, music, and poetry like the living human artist.

In Mary Shelley's novel *Frankenstein*, a young scientist fantasises about creating a real-life human being. He works relentlessly for years collecting and putting together the most beautiful parts of corpses that he can find. But when Frankenstein's monster finally comes to life, Dr. Frankenstein is horrified:

> I collected the instruments of life around me, that I might infuse a spark of being into the lifeless thing that lay at my feet (....) I saw the dull yellow eye of the creature open (...) how can I describe my emotions at the catastrophe, or how delineate the wretch whom with such infinite pains and care I had endeavoured to form? His limbs were in proportion, and I had selected his features as beautiful. Beautiful! – Great God! His yellow skin scarcely covered the work of muscles and arteries beneath; his hair was a lustrous black and flowing; his teeth of pearly whiteness; but these luxuriances only formed a horrid contrast with his watery eyes, that seemed almost of the same colour (...) The different accidents of life are not so changeable as the feelings of human nature. I had worked hard for nearly two years, for the sole purpose of infusing life into an inanimate body (....); but now that I had finished the beauty of the dream vanished, and breathless horror and disgust filled my heart.[30]

Feelings of affection towards an inanimate object with human-like features may quickly turn into disgust. In the horror movie series *Child's Play*, every kid's dream doll, the "Good Guy", turns into the nightmare killer doll, Chucky, when infused with the soul of a serial killer. We pass shop windows daily with mannequins without giving them an extra thought. However, countless horror movies touch upon our inner uneasiness with the creepy human likeness of these life-size dolls when bringing them to life or using them as murderers' props.

Should a machine be designed human-like?

When the British science fiction author Arthur C. Clarke visited the Bell Labs in the US, among others housing Claude Shannon, the father of information theory, his encounter with Shannon's "ultimate machine" left him with an "eerie" feeling:

> Nothing could look simpler. It is merely a small wooden casket the size and shape of a cigar-box, with a single switch on one face. When you throw the switch, there is an angry, purposeful buzzing. The lid slowly rises, and from beneath it emerges a hand. The hand reaches down, turns the switch off, and retreats into the box. With the finality of a closing coffin, the lid snaps shut, the buzzing ceases, and peace reigns once more. The psychological effect, if you do not know what to expect, is devastating. There is something unspeakably sinister about a machine that does nothing – absolutely nothing – except switch itself off.[31]

"The Uncanny Valley" is the space of eerie feelings that humans sometimes enter when encountering a robot or another device designed with human-like characteristics. The concept was described in a short article by the Japanese roboticist Masahiro Mori in 1970.[32] He writes that a human being's first reaction to a robot or device (such as a prosthetic limb) with human likeness is "affinity", a kind of sympathy, precisely due to the human likeness. Nevertheless, the moment we realise the artificiality of the thing, we experience an "eerie sensation". This is when we have entered the "Uncanny Valley" of the "living dead", a space so uncannily close to the "Still Valley" of the "corpse", the ultimate human moment of death when we stop moving. He finds it very difficult to identify that exact moment when the feeling of affinity transforms into eeriness and therefore believes that the safest way forward in robot design is to pursue "nonhuman design" deliberately:

> To illustrate the principle, consider eyeglasses. Eyeglasses do not resemble real eyeballs, but one could say that their design has created a charming pair of new eyes. So we should follow the same principle in designing prosthetic hands. In doing so, instead of pitiful looking realistic hands, stylish ones would likely become fashionable.

How is the inception of an AI system different from the birth of a human child?

We will never stop trying to seize life; this is human nature, but human life cannot be captured. Sanders only caught a glimpse of the lives of the German population in his thousands of photographs; the Greek wine kylixes have scenic, often comic, depictions of life painted on them, but only the broken edges of the cups know the mouth that touched them to drink.

William Shakespeare's Macbeth seems desperate in his existential turmoil when he bellows:

> Life's but a walking shadow, a poor player, that struts and frets his hour upon the stage, And then is heard no more. It is a tale Told by an idiot, full of sound and fury, Signifying nothing.[33]

I grew up with the echo of this quote in my mind. My mother, an English teacher in the Danish Air Force, repeated it so many times that I, as a very young child, had already memorised it and would say it out loud to the amazement of my teachers (little did they know that this was the only Shakespeare quote I knew). I always felt a surge of discomfort about these gloomy words, though. When I later read Macbeth in high school, the darkness and weight of that statement didn't leave me. Why can't we catch that fleeting shade and hold it still? We may get a glimpse of human life in a passing moment; we may intuitively sense it, but we cannot grip and take hold of it. Life always escapes. I felt Macbeth's burden.

Later in life, I was introduced to the books and papers of the "vitalist" Henri Bergson, and it was through reading his works that my feelings about the fleeting nature of human life changed. I have and will be revisiting Bergson and his views on humanity, life, and the human being throughout this book. In fact, with the vitalists of the late 19th century and Bergson, we see that life as a defining character of humanity is a human power precisely because it defies representation and simplification and escapes efforts to hold it still. So, I encourage you to do what I did then: change your perspective on human life and take a different, less dim look at the power of it. Is it not this transient character of life that makes it special? Its open-endedness, uncertainty, and resistance to immobility that is the true potential of human life. And do we today not need to make a more serious effort to protect this trait of human power from that which wants to hold it still?

Today, when the reality of human life is, in truth, one that predominantly seeks to trap life in data representations and classify and categorise it in algorithmic systems, I believe that Bergson's ultimate mission is more critical than ever. Remember Chapter 1 on Creativity and how he defined evolution? What he wanted to do was to change not just a scientific (to replace mechanistic scientific approaches to "life" and "consciousness"), psychological (to escape habits that trap new experiences in old[34]) and philosophical (to challenge metaphysical materialism) paradigms, but also to transform the social practices and political processes of the time (as he tried to do with his significant contributions to international peace politics). He wanted to guide a societal evolution based on creative principles, rather than mechanical – towards an "open" society, away from a "closed" one.[35] Regrettably, despite the promises for greater openness and global human connection of the late 20th century World Wide Web cyber-libertarian movements, it seems that we are today in the 21st inhabitants of the "closed societies" that Bergson worried about.

Is AI an existential threat to human life? In what way?

In early 2023, hundreds of prominent business people and other renowned figures released a chilling public statement on AI, including OpenAI's CEO Sam Altman, Google Deepmind's Demis Hassabis and co-founder of Microsoft Bill Gates.[36] It consisted of only one line:

> Mitigating the risk of extinction from AI should be a global priority alongside other societal-scale risks such as pandemics and nuclear war.

Should we be concerned? Is AI an existential threat to humankind on par with such disasters? If we try to answer the question with Bergson, we can confidently say that it is not the extinction of our species, we should be fearing from these living dead AI systems. At least not in any traditional sense. In fact, we might even be wasting precious time worrying about, listening to and following the leads of tech gurus with unrealistic claims about the power of AI to pose "existential threats" and cause "human extinction". Nevertheless, according to Bergson's evolutionary theory, there are other ways of stagnating a species' evolution that we want to avoid.[37] What we should really fear and consider is AI's impact on the human creative life drive, because a loss of creativity and ability to adapt to change is also a potential end of a species' evolution.

NOTES

1. *The Cold Gaze – Germany in the 1920s*, Louisiana, 2023.
2. Bergson, H. (1914) *Creative Evolution* (translated by Arthur Mitchell) (p. viii), Macmillan and Co. (originally published in 1907).
3. Vaughan, M. (2007) "Introduction: Henri Bergson's "Creative Evolution", *Substance*, 36(3), 114: *Henri Bergson's "Creative Evolution" 100 Years Later (2007)* (pp. 7–24), The Johns Hopkins University Press.
4. Boccioni, U. (1910) *Technical Manifesto of Futurist Painting*, https://www.arthistoryproject.com/artists/umberto-boccioni/technical-manifesto-of-futurist-painting/. Sadly, the futurist artistic movement of which Boccioni was a key figure became closely associated with the facist transformation of Italy. Quite the opposite of the Jewish philosopher Henri Bergson, who among others, worked closely with US President Woodrow Wilson's administration to establish the predecessor of the United Nations, the international peacekeeping organisation the League of Nations.
5. Quoted in Mather, D. S. (2023) "Chromatic Futurism Vitalizing Painting, Sculpture, Music and Life's Energies" in *Vitalist Modernism*, Routledge.
6. Vaughan (2007).
7. Bergson (1914/1907).
8. Alpaydin, E. (2016) *Machine Learning*, MIT Press.
9. Bowker, G. C., Star, S. L. (2000) *Sorting Things Out: Classification and Its Consequences*. Inside Technology, MIT Press.
10. Chiusi, F., Fischer, S., Kayser-Bril, N., Spielkamp, M. (Eds.) *Automating Society 2020*, Algorithmwatch, https://automatingsociety.algorithmwatch.org/
11. Pasquale, F.A. (2015) *The Black Box Society – The Secret Algorithms That Control Money and Information*, Harvard University Press.
12. Translated from Danish: "Jeg saa dig blusse og jeg skjalv Forstillelsen kun skjulte halv En Attraa i vort Bryst Som aldrig Dyden veed at skatte For hvilken Ord er altfor matte En ubegrændset Lyst", Aarestrup, E. "Jeg saa dig blusse og jeg skjalv" in Samlede Skrifter, Brix H. (red.) (1976), Dansk Sprog- og Litteraturselskab, C.A. Reitzels Boghandel, Kbh. (p. 149), https://kalliope.org/da/text/aarestrupudata4
13. Translated from Danish: "Hold fastere omkring mig Med dine runde Arme; Hold fast, imens dit Hjerte Endnu har Blod og Varme. Om lidt, saa er vi skilt ad, Som Bærrene paa Hækken; Om lidt, er vi forsvundne, Som Boblerne i Bækken", Aarestrup, E. (1838) "Angst" in Digte, C. A. Reitzels Forlag (p. 268). https://kalliope.org/da/text/aarestrup1838b39
14. See e.g. Johanne Louise Heiberg's depiction of a conversation about Emil Aarestrup between her renowned cultural critic and author husband Johan Ludvig Holberg and one of Aarestrups followers: "The poet Aarestrup had by this time come into fashion, and Mrs. Drewsen was one of his followers. Heiberg now teased her to make fun of him, for he could not find taste in the creations of his muse (...) When she despaired over this, Heiberg improvised a poem which, to make it even more parodic, he gave the melodious title "à l'Aarestrup". And now he recited his Impromptu with

comic pathos (…) This Impromptu was of course rewarded with laughter from us all, and thus ended the battle for Aarestrup." (Translated from Danish from Heiberg, J.L. "Et liv gjenoplevet i erindringen af johanne luise heiberg, 4. Reviderede udgave ved aage friis under medvirken af Elisabeth Hude, Robert Neiiendam og just rahbek", I. Bind, første del, 1812–42, Gyldendalske boghandel, Nordisk Forlag, 1944, https://tekster.kb.dk/text/adl-texts-heibfr05val-root

15. Translated from Danish: "En vis Liderlighed er den eneste Gehalt (indhold) i disse Poesier", in S.S. Mittet, *Emil Aarestrup*, Litteratursiden.

16. Weber, M. (2001) *The Protestant Ethic and the Spirit of Capitalism* (introduction by A. Giddens), Routledge (original translation from German by T. Parsons 1930, original German text 1905).

17. Giddens in Weber (2001/1905, p. xii).

18. Weber (2001/1905, p. 123).

19. Akmenkalns, J., Sneed, D. (2018, June 18th) *The Symposium in Ancient Greek Society*, University of Colorado https://www.colorado.edu/classics/2018/06/18/symposium-ancient-greek-society

20. Bloom, P. (2022, January 23rd), "Hedonism is overrated – to make the best of life there must be pain, says this Yale professor", *The Guardian*. https://www.theguardian.com/lifeandstyle/2022/jan/23/hedonism-is-overrated-to-make-the-best-of-life-there-must-be-pain-says-yale-professor

21. Onfray, M. (2015) *A Hedonist Manifesto the Power to Exist* (p. 104), Colombia University Press.

22. Onfray (2015, p. 62).

23. Onfray (2015, p. 164).

24. Weber (2001/1905, p. 60).

25. Hertz, N. (2020) *The Lonely Century: How to Restore Human Connection in a World That's Pulling Apart* (p. 73), Sceptre.

26. Rao, L. (2011, July 3rd) *Sexual Activity Tracked by Fitbit Shows Up in Google Search Results*, http://techcrunch.com/2011/07/03/sexual-activity-tracked-by-fitbit-shows-up-in-google-search-results/

27. Cha, A. E. (2015, May 19th) *Health and data: Can Digital Fitness Monitors Revolutionise Our Lives?*, https://www.theguardian.com/society/2015/may/19/digital-fitness-technology-data-heath-medicine

28. Kurama, V. (2022, October 11th) *What Is Collaborative Filtering: A Simple Introduction How Recommender Systems Use Collaborative Filtering*, https://builtin.com/data-science/collaborative-filtering-recommender-system

29. Williams A. (2024, February 14th) *When Love and the Algorithm Don't Mix*, https://time.com/6694129/dating-app-inequality-essay/

30. Shelley, M. (1989) *Frankenstein* (p. 66), Puffin Books (originally published 1818). Thank you Clara for reading Frankenstein and refreshing my memory with this quote that you showed me you had underlined in your book.

31. Clarke, A. C. (August 1958) *The Ultimate Machine*, Harpers.

32. Mori, M. (2012) "The uncanny valley" (originally published in Japanese 1970) (translated by K. F. MacDorman and N. Kageki), *IEEE Spectrum*, https://spectrum.ieee.org/the-uncanny-valley#_ftn1

33. Shakespeare, W. (1984) *Macbeth* (edited by Kenneth Muir), Methuen Drama (originally published 1623).
34. Vaughan (2007, p. 12).
35. Bergson, H. (1977) *Two Sources of Morality and Religion* (translated by A. Audra & C. Brereton), University of Notre Dame Press (originally published in French, 1932).
36. "Statement on AI Risk", Center for AI Safety, https://www.safe.ai/work/statement-on-ai-risk#open-letter
37. Vaughan (2007).

CHAPTER 4

Intuition

How Does a Machine Make a Decision?

A young French man came to his philosophy professor for advice on an ethical dilemma of his. This was during the Second World War, and his brother had been killed as a soldier in a German attack. The young man now wanted to revenge his brother by going to England and joining the Free French forces. But he was torn. His mother, who was already grieving the loss of her firstborn son, would undoubtedly suffer great pain with the prospect of potentially losing her other son. What should he do? How should he weigh his options? Jean-Paul Sartre, because this was the professor the student had asked the question to, answered his student:

> If values are uncertain, if they are still too abstract to determine the particular, concrete case under consideration, nothing remains but to trust in our instincts.[1]

We will return to Sartre and his young student, but first, I want you to consider what Sartre is asking this young man to do. If I am not wrong, your first thought will probably be that it's odd to tell a young person to trust in their instinct when they are asking for advice on an ethical dilemma that will profoundly affect the course of life; that might

DOI: 10.1201/9781003527855-6

even lead to the early end of this young life. How can one make such a grave decision based on this – an inexplicable inner feeling? How will he ever be able to make this decision without any guidance apart from a "gut feeling"?

We often see human intuition depicted in movies as a mystical force, a "gut feeling", of specially gifted individual human beings. In M. Night Shyamalan's thriller *The Sixth Sense* from 1999, a boy can see and communicate with the dead. He has an eerie intuition, a "sixth sense", providing access to a parallel invisible world of ghosts. In Lana and Lily Wachowski's science fiction movie, *The Matrix*, from the same year, the main character, Neo, also hinges on his inner instincts, something that feels more "real" than the virtual world he is navigating in.

If we think of intuition like this, as mystical hunches of humans with supernatural powers that are impossible to explain, it seems ludicrous to put one's blind faith in something like that and use it as the basis of critical decision-making. One could argue that in the "real world", we need more than this. To reach "real" valid conclusions and "truths", we need evidence, arguments, and information to weigh in our options when deciding about something like life and death. Thus, it is evident that a mystical and highly subjective intuition is not just strange; it is "anti-scientific", and it should not be the basis of critical decision-making.

This is at least how the Nobel Prize-winning psychologist Daniel Kahneman described human intuition. Human intuition is not a "sixth sense". It is not magic, he argued. Intuition is nothing more than information processing done very fast.[2] This is also why experts have better intuitions than non-experts. They simply have more expert information to process and make a more qualified quick judgement than non-experts. The human intuitive decision-making, when we "think fast", as he called it, is thus also a weakness. It belongs to what he called our "system 1", which allows humans to think and act fast, make effortless quick associations, judgements and predictions, and respond to recognisable patterns based on past experiences. For example, when a cat runs across the street, and you quickly predict the danger and halt the car.

Of course, if that is all human intuition is, then when we process information too fast and make quick judgements based on our past experiences, these will always, as Kahneman argued, be more biased than when we "think slow", process information slower and apply a critical, logical analysis. With this slower "system 2", we make more of a thinking effort, apply careful analysis and predictions based on reasoning, weighing in logical options, evidence, and critical reflection.

Is the analysis, the data, and the criteria for an AI system's information processing the same as human intuition?

Evidently, Kahneman believed we should be more modest about what we are capable of and better at "disciplining" our fast, intuitive assessments. What seemed to be an echo of early Cybernetic theories, he and his co-authors of the widely recognised book *Noise: A Flaw in Human Judgment* described the human mind as a type of machine and human judgement as just another form of "(…) measurement in which the instrument is a human mind".[3] This is why he also believed that human judgement is greatly inferior compared to algorithms. Algorithms can help create a much better "decision hygiene" as "the goal of judgment is accuracy, not individual expression".[4] These accurate algorithms remedy "(…) the sheer magnitude of system noise and the amount of damage it does".[5] The "damage" that our human flawed judgement represents in society, that is. Replacing human judgement with algorithms or just generally being more "guided" by algorithms, Kahneman saw as a big improvement in society. In fact, there are no limits to the positive prospects of building organisations and a society that strives to "reduce noise" and to achieve "a less noisy world". It would, as he stated, "(…) save a great deal of money, improve public safety and health, increase fairness, and prevent many avoidable errors".[6]

If making judgements was only about this, being "accurate", then reducing the noise with an algorithm would be no problem. We could rely on algorithms mostly to hire the accurate candidate for a job or decide the correct sentence for an offender. Intuition would certainly be useless in such a clear-cut reality of things and its applicable scoring card. Of course, we would first need to translate the world, society and humans into something much less complex. After all, we would need to fit them into the classification scheme and criteria of the algorithm. For example, connections between people would not be relevant when hiring a person. All that would matter is whether that person is "right" or "accurate". Showing or not showing remorse would not be meaningful when giving a sentence because how do you translate remorse into an algorithm? And, of course, we would need to collect data, heaps of it, to feed those algorithms and ensure that we are all accurate and fit into the various systems we are judged in and judged by every day.

Is an AI system biased? Where does the bias come from?
Can an AI system be non-biased?
Does an AI system discriminate? Does it care?

Throughout history, human intuition has been either praised as a hidden human potential, a mystical, spiritual, subjective capacity of individual human beings, or it has been like Kahneman did, criticised for precisely that: human intuition is not magic; it is a system that makes an otherwise slow human being capable of responding quickly to a fast-changing reality. However, it is also a human bias in disguise, our internalisation of social order, and a form of anti-scientific subjective human flaws that must be tamed. Kahneman believed in the power of the algorithm, and he thought that our aversion towards algorithms and machines is painted by our sympathy towards the natural, the human. He stated:

> The story of a child dying because an algorithm made a mistake is more poignant than the story of the same tragedy occurring as a result of human error, and the difference in emotional intensity is readily translated into a moral preference.[7]

While there is a part truth to this observation – humans prefer the familiar, identification is a solid human appeal – I also believe there is a lot more to it than that.

Let's for a moment imagine two different processes that, although leading to the same result, a child dying, are very different. A machine learning system applied in a healthcare situation analyses real-time patient data collected from various sources, such as symptoms and vital signs, like temperature, oxygen levels, blood pressure, and health records, which have been translated into interoperable data. It first pre-processes the raw data by cleaning it, ensuring its quality. Then, it analyses it and makes decisions based on patterns learned from historical data. It makes a decision that is then implemented. This time, it is an erroneous one. The data didn't add up. Now, imagine a different process where human beings, memories, and histories are intertwined. A team of medical practitioners, not without technical means of course, this is not in question, they do have access to technology that can support them in their work (even algorithmic tools that can measure and analyse symptoms and vital signs). But what we want to focus on here is something other than the processing of data. Think about the feeling of the life of a child, the desperation of the relatives, the burden or the backing of a hospital system, the culture of the organisation, the medical team's despair and compassion. Perhaps also efforts to understand the social and ethical dimensions of the situation. There are interchanges with parents, colleagues, and other experts. Decisions are made. This time, they were also erroneous.

If we now think of these situations as not just particular situations but as images of society and then ask – which kind of society would you prefer? Which qualities of the societal processes do you value? Most people would likely again say that they preferred the "human" one to the "algorithmic" one. But this time, it is not just because they have more sympathy towards humans or feel more familiar with nature. We would prefer this because the situation and the process that leads to the child's death are privileging humane qualities. Of course, I am not arguing here that the algorithmic analysis is not valuable in a healthcare-critical situation, but please consider the dominant character of how we do things. Do we prioritise more humane practices? Do we appreciate human intuition as a different quality of a process?

Now, if we were to consider intuition as more than just a quick way of processing information, which I think it is, what would a "more humane" intuitive way of doing things mean? Let me explain with a different example.

Alfredo, the man who runs the film projector of the cinema in the 1988 film *Cinema Paradiso*, loses his eyesight in a fire at the cinema. Nevertheless, to the amazement of the boy Toto, who is now running the film projector, he can still sense when the image is blurry. Later in life, when the girl Toto is in love with appears in a film he made himself, the blind Alfredo exclaims excitedly: "It's a woman", although he cannot see what is happening on the screen. In these two scenes, Alfredo has an intuition that amazes the people around him and *Cinema Paradiso*'s viewers. He extra-ordinarily senses something that would not usu-ally be possible for someone who cannot see. What does these scenes' depiction of human intuition tell us? How is Alfredo's intuition differ-ent from Kahneman's complex logical algorithms? Here, I want you to think of human intuition as something different, not the opposite, just different from the analysis of an algorithm. I am not asking you to think of human intuition as a mystical, inexplicably superpower. Instead, I would like you to think of it as the humane quality of a situation and a process.

Thus, first, you must understand that *Cinema Paradiso* is a film celebrating human power. The filmmaker Giuseppe Tornatore's love for human art in film and the cinema is profound; the affectionate depiction of the village people whose lives we see evolving inside and around the cinema from the priest that censors the kisses on the screen to the "crazy" man doing his rounds outside on the square with the frantic repetition "La piazza mia!" With much empathy, he portrays the human relations and emotional bonds between people: Toto and his mother, Alfredo and Toto, and Toto and the girl he loves. Also, the lurking brutality of the type of human power that destroys and kills is present in the movie. Alfredo somewhat replaces Toto's father, who was killed in the war. The priest is censoring what people can see in the cinema according to the dogma of the Catholic church. In a depiction like that, intuition is undoubtedly more than just the expert brain processing information when detecting sound clues from a film projector or physical clues, such as the sweat on the skin of a boy in love. Maybe Alfredo sensed the boy Toto's feelings towards the girl precisely because of his love for the boy? Did he hear the familiar murmur of the village people in the cinema when the screen blurred? Did not all these emotions, immediate senses, past and present experiences, cultural knowledge, and also modes of a human brain's information processing come together in an intuition that allowed Alfredo to "see" the malfunctioning of a projector and to act with thrill, excitement, and compassion towards the young boy next to him when he realised that he was in love? Now, I ask you again not to assess the success rate of the logical analysis of an algorithm compared to that of a human intuition but to think about which kind of society you want to live in. One that admires and celebrates the complexity of human intuition or one that seeks to tame it?

How many different capacities does an AI system's decision-making consist of?

Thus, let's change our perspective on human "intuition" from one that considers it mainly a human urge to be disciplined to one concerned with the qualities of human power. Let's think of intuition as a human capability consisting of a particular approach, not contrary to a rational scientific one, just one different. The best place to start an exploration, as such, is to look at the role of intuition in the life and work of an artist.

The music composer Marianna Filippi told me she uses her intuition in very concrete ways in her work. In fact, she sees intuition as fundamental to her creative work composing music as this is the only way she can give shape to her emotions in music. There is no other way. Other more mechanistic approaches would limit her, she told me:

> The way I work is entirely intuitive. I use my intuition to express the emotions that I can't express in words. It is like a power to transmute my emotions into something tangible without putting any restraints on myself, without trying to put myself into some box. It is like the truest state of being, a state of mind where I'm just immediate. I'm not thinking. I'm just trusting myself to be able to create. I trust what I feel is the right thing to do. I am creating something new by myself. It brings me to a place where I stop being influenced by external things. I'm just improvising and coming up with things I immediately think about. It is very different from, you know, sitting down for hours, maybe even weeks, meticulously planning a piece. Follow this exact rule, these exact things, this very structure. I don't want to sit there and write some structure to follow because I never actually end up following it. All the times that I've tried that, it never works. I want to listen to my immediate thoughts. I use my intuition, and I'm in the moment, not thinking about some other structure that was written before.[8]

With the composer and her approach to music in mind, we may now consider intuition not just a "lesser" form of a better rational form of thinking based on predefined rules and programmes. Instead, it is a different way of being in the world, making decisions, and creating things.

What capacity does an AI music generator use to create music? Does intuition play a role?

While still considering human intuition a form of information processing, the psychologist Carl Jung didn't see it as just a lesser, more slack way of processing information. He associated intuition with a function of a particular human personality type that grasps and approaches the world and personal and social relations in a specific manner and acts accordingly. He believed we can recognise different personality types by the predominance of specific inclinations, interests, and attitudes in the world and towards others.

Famously, Jung described the "introvert" and "extrovert" personality types, but he also presented four specific prevalent functions of these: "thinking", "feeling", "sensation", and "intuition". As dominant modes of understanding and relating to the inner and the outer world and the things and people in them, the "thinking function", for example, is mainly concerned with logic, rational analysis, and objective assessment. With a prevalent "intuition" function, on the other hand, you are more inclined towards the abstract to notice hidden patterns and meanings that one does not immediately see. You are also more at ease with uncertainty and exploring things beyond what is immediately given and considered facts of reality. Generally, Jung describes this personality function as one that will result in creative processes and innovation. Of course, Jung didn't believe in "pure" types. No human being is the complete expression of one personality type. We are a complex mixture of these with inclinations towards some qualities rather than others. What is important to note here is that Jung does not perceive intuition as a "lesser" form of rational "thinking". It is just a different form.

How does an AI system's decision-making capabilities differ from human intuition?

Building on this, we may now understand human intuition as something more than giving in to whatever irrational feeling we may have in the spur of the moment. Instead, we should consider it a different inclination and way of approaching things.

Henri Bergson described intuition as a different effort than the effort of the intellect. He believed humans are limited by an intellect that represents reality by dissecting, artificially reconstructing, and analysing it with "ready-made" concepts and ideas. The intellect is also primarily concerned with what is useful to the task at hand and, therefore, is mainly concerned with the present moment. The past and the future are only there to support the present.

How does AI treat time? What role do the historical training data (the past), the data processing (the present), and the predictive analysis (the future) play?

Intuition, on the other hand, is an effort to do something different: to perceive uninterrupted time in all its totality, to place oneself in a moving and changing reality and the plurality and diversity of it.[9] Bergson used time as an example of these two very different efforts. While the "clock time" of the intellect makes time manageable by dissecting it into 1–12 numbers, intuition is an attempt to perceive the "duration" of time as it passes. According to Bergson, this task is difficult, as our intellect shapes our approach to the world. Yet, we must try to surpass our intellectual limits:

> It is more than human to grasp what is happening in the interval. But philosophy can only be an effort to transcend the human condition.[10]

In 2024, when speaking with Professor Renuka Singh, the editor of six of the 14th Dalai Lama, Tenzin Gyatso's books, I recognised a similar engagement with the intuitive state of being and approach. We talked about the role of what she referred to as an "emotional approach" to the world (see also Chapter 2) and about human "wisdom" and making "wise decisions" (see also Chapter 7). She said:

> It's a state where you are living in the moment. And you can see and understand the reality in its totality. There's no distortion of reality, there are no illusions left. At that moment, you act in a very wise manner. Then you do the best possible for yourself and others.

Renuka Singh explained that according to Tibetan Buddhism, the universe holds all knowledge. Humans are mere receivers of knowledge that exists like vibrations in the universe. We do not inherently hold knowledge, but our minds can tune into the knowledge of the universe. For instance, when a person enters a meditative state, that person can tap into this frequency and receive knowledge, she said:

> This is how we discover and become creative by being receptive. And it happens only when you have that intuitive wisdom. There is a rational level, and we are all operating at that level, but not everyone can intuitively receive the knowledge coming from elsewhere.

Renuka Singh also told me that as a student she went to what she called a "radical" university, so I asked her about her "activism" as a student. At first, she said she was "never an activist (…)", then she retracted that statement: "(…) but in philosophy yes". She continued explaining the "emotional approach" of Buddhism, while at the same time, she seemed to want to assure me that this was also a very "rational", practical, that is, approach:

> You have to work very logically towards understanding reality. You have to understand the emptiness of all the phenomena that exist around you in a very rational way. That they are not independent, they are interdependent. Reality is empty of inherent existence. When you grasp the emptiness in a logical way, often through meditation, your attitude towards the world and everything changes. Your mind gets transformed from the negative to the positive. You become forgiving. You become universal. You become a humanitarian. You have compassion. You have empathy.

Bergson considered the intellect's "rational" approach necessary when dealing with concrete, tangible things ("matter"). But can we also in the same way approach and create things constructively with intuition? Does intuition have any practical role in a society beyond a purely "philosophical" one? I want to claim that we should be able to do so. Renuka Sing is indicating that this is so. We have also seen how the composer Marianna Filippi uses her intuition as a very concrete method for creating music.

While Bergson was less inclined to describe intuition as a practical approach, the philosopher Gilles Deleuze, in his book *Bergsonism*, described "Intuition as Method".[11] This is an approach, he argues, that is not defined by a predefined goal but constantly moves towards an undefined future state of "becoming". It allows perceivers to immerse themselves in the qualitative and temporal context of a situation, considering its conditions as a totality, over time, rather than focusing solely on the present moment.

To explain this, we could also look at intuition's role in sociological theory. Georg Simmel, the scholar who defined sociology as a discipline, correlated intuition with the inductive "bottom-up" method in sociology, where the researcher uses individual data to develop general theories rather than approaching the field of study with a ready-made concept and idea fitting it into the pre-defined schema. He believed it is fundamental to our understanding of the world and society to try to understand the whole from the one, not only to differentiate or crystalise what we already know.

Because every little part of the world and society influences each other and are thus intrinsically intertwined:

> We could indeed not call the world one if each of its parts did not somehow influence every other, if anywhere the reciprocity of the influences, however mediated, were cut off.[12]

This thinking transcends into Simmel's perception of time as well. From our previous discussion in Chapter 2 of feeling and emotion, you might recall that he also conceived of time in this way: life and death as inseparable. Living is also dying, he said. Time is duration, and intuition is an effort to place oneself within an everchanging duration of time in which every state of being is simultaneously transitioning into another state.

The anthropologist Tim Ingold describes intuition as an "immersive experienced" research approach. He introduces the concept of "dwelling", the very immersion of the researcher in the environment he studies, suggesting that dwelling is a way of being and seeing our surroundings that allows us to perceive them from everywhere simultaneously.[13]

This is also the way of the troll creature Dunderklumpen in the Swedish story and cartoon film released in 1974 by poet and author Beppe Wolgers. As Dunderklumpen sings heartedly in its little song about itself:

> And myself, I play on the stumps and dance 1, 2, 3, 4. I'm forty-two years old, but I remember when I was two, ten, and thirty-two years old. Have you understood? Well, not that. Well, I mean that (…)

Dunderklumpen can whistle like a three-year-old and break a branch like a 30-year-old, shoot a sling slot like a ten-year-old and shout "Take cover, take cover" like a 40-year-old, the troll explains to its animated toy friends and then exclaims excitedly: "I'm still everybody I once was". This troll is timeless and ageless; what is right here and now doesn't matter. Dunderklumpen understands and perceives every time and every place it has been to – not consecutively and exclusively limited to the present moment and utility of that moment, with a perspective bound in space and time – but all at once. Travelling through magic forests and over mountains, meeting creatures in all shapes and forms on its way, it is, above all, curious and open-minded when approaching everyone it meets with wonder and appreciation. In effect, Dunderklumpen relates to its animated toy friends and other creatures in a spontaneous, child-like manner and with an intuitive understanding of their needs and individual uniqueness.

To Dunderklumpen's creator, Beppe Wolgers, who also played Pippi's pirate father in the Pippi Longstocking TV series, being called "childish" is a compliment. He was a proponent of the child's perspective, their mysterious world and way of understanding and approaching their surroundings, which he described as an increasingly alien approach in an adult world of rational restraints, dogmas, and political, economic, and social demands.[14]

Wolgers is not the first to see the value of children's intuitive and playful approach to the world. Most famously, the Italian physician and educator Maria Montessori developed a philosophy of education in which children's intuition and curiosity are encouraged through exploration, independence, and hands-on approaches to learning. She believed that children were not just lesser evolved adults but were "endowed with a special psychic nature", a "real, constructive energy", "alive", and "dynamic".[15] To illustrate the power of the child's approach, she uses the example of learning a new language. While children have infinite capacities to acquire and make new languages their own, adults will never be able to learn a language in the same way. Hence, valuing the child's world and approach is something we need to make a goal when developing our civilisations as it means at the same time cultivating a hidden potential of humanity.[16]

Indeed, Marianna Filippi, whom I spoke with about her methods when creating musical pieces, emphasised the playful character of the improvisations, she initiates a music composition process with, as "childlike":

> When I begin a piece, I usually use improvisation, a kind of playfulness. I use my intuition to just play around with things. It's like returning to a whimsical kind of childlike state of mind.

With this "hidden human potential" in mind, we could think of the childish Dunderklumpen and the Anthropologist Tim Ingold as representatives of essential modes of relating to and exploring the world. What they do when visiting an environment, whether it be magic forests and their inhabitants or the Saami people in Northern Finland, is to dwell in those places. They do not try to make sense of them from a detached "outside" position, applying whatever premade ideas and dogmas they possess about that environment and its people and cultures. They use their intuition to place themselves within these environments to understand them on their own terms.

Can AI "dwell" in an environment? How does AI approach its environment? Can AI be "childish"?

An intuitive approach as such, one could also think of as sympathetic. This is how Henri Bergson describes intuition:

> By intuition is meant the kind of ·intellectual sympathy by which one places oneself within an object in order to coincide with what is unique in it and consequently inexpressible. Analysis, on the contrary, is the operation.[17]

Here, we want to try to understand human intuition with our previous description of the power of human feeling and emotion, which is, of course, at the same time, our greatest potential and greatest weakness. We don't only have emotions of compassion; we are also fearful, angry, and envious, and these negative emotions are the roots of many of the biases that sieve into and shape personal decision-making and the politics of our societies. For example, we can trace fears of what we don't know and are unfamiliar with (what we can't identify with) in nationalist politics and the tightening of securitisation or immigration policies.

Having said that the kind of intuitive sympathy Bergson and others refer to comprises a different way of relating to the world than the rational, logical way of the "intellect". The intellect works with symbolic representation, discursive and symbolic knowledge, and thus, as Bergson calls it, "artificial reconstruction". The problem is here that when doing so, it is working with "the shadow alone", a "clumsy imitation" that always "satisfies an interest".[18]

Ingold's dwelling in the cultures that he studies and Dunderklumpen's childish, curious, and open exploration of magical forests, on the other hand, constitute entirely different ways of relating to things and people. Intuition is, in essence, not a less valuable way of perceiving, interacting, and exploring; it is just a different kind of intellectual, more sympathetic, or even emphatic, effort. As Henri Bergson describes it:

> (...) our intelligence can follow the oppositive method. It can place itself within the mobile reality; and changing direction; in short, can grasp it by means of that intellectual sympathy, which we call intuition.[19]

Does a machine have the capacity to approach the world with empathy? Can AI function without data (representation)?

There are countless examples of influential humans in history exhibiting this kind of intuitive approach, or "intellectual sympathy", to engage not only in the arts but also in politics and science.

The Indian civil rights leader Mahatma Gandhi is an excellent example of a person who advocated nonviolent resistance to British colonial rule, stressing the importance of listening to one's "inner voice", or as he called it, a sort of "inner prompt" when making decisions.[20] It is not surprising to see this in a person like Gandhi. But, there are also other examples of people that you would not normally associate with an approach as such.

The theoretical physicist Albert Einstein, who spent all his life developing and discovering the logical laws of physics, believed in the power of human intuition and "creative thought" as the key source of scientific discovery.[21] Intuition and creativity, he and the physicist Leopold Infeld write in their book *The Evolution of Physics*, enable scientists to think beyond the most obvious. In fact, it empowers them to see beyond obvious explanations, following the clues, as they say, that are not immediately given. They argued that an intuitive approach enables non-linear thinking, leading to innovation and breakthroughs out of the ordinary:

> To raise new questions, new possibilities, to regard old problems from new angles. Requires creative imagination and marks real advance in science.[22]

Solutions to scientific problems Einstein and Infeld saw as a collection and arrangement of facts that could be made coherent and understandable with intuition and creative thought.[23] It is like solving a mystery, they argued. The challenge is to use one's intuition creatively, that is, by not falling into the trap of following only the most obvious clues:

> In a good mystery story the most obvious clues often lead to the wrong suspects. In our attempts to understand the laws of nature we find, similarly, that the most obvious intuitive explanation is often the wrong one.[24]

In narrative fiction, we often encounter detective characters with exceptional intuitive abilities in solving mystery crimes. Most famously,

Sir Arthur Conan Doyle's private detective, Sherlock Holmes, solves one seemingly impossible crime mystery after another with a logical deductive intuition. The narrator of the detective stories and Holmes' only friend, Doctor Watson, describes this as more than just a method the detective applies. It is an urge and a defining character of his human genius:

> Then it was that the lust of the chase would suddenly come upon him and that his brilliant reasoning power would rise to the level of intuition, until those who were unacquainted with his methods would look askance at him as on a man whose knowledge was not that of other mortals.[25]

Holmes' intuition is analytical and rational. He observes facts and discovers clues based on reason and empirical evidence. What at first appears mysterious and supernatural, he always manages to dissect into a very rational explanation. The Hounds of Baskerville are not ghost demon dogs, as the frightened people of the village believe they are, but very real dogs released by a man plotting to kill his family relations to inherit a family estate.

While Holmes is described as intuitive, he is not an emotional man, not the slightest bit interested in feelings and emotions, in his cases' broader social or ethical implications, nor is he interested in the relation with other human beings. He only appears to have an emotional life world conveyed through the music he plays on his violin. In all other areas, he is ruthless, cold, and emotionless, very much like his mortal enemy, the criminal mastermind Professor Moriarty. Thus, Holmes is also depicted by Watson as incapable of even loving the only woman in his life, Irene Adler:

> All emotions, and that one particularly, were abhorrent to his cold, precise but admirably balanced mind. He was, I take it, the most perfect reasoning and observing machine that the world has seen.[26]

Commissario Guido Brunetti, the Italian detective from the *Commissario Guido Brunetti* series of books by Donna Leon, is a different kind of man. He is not as interested in the game of putting clues together to solve the crime as he is in understanding the social implications, the motives behind the crimes, and the lives and emotions of the people he engages with. Brunetti is an educated man from one of the world's cities with the richest human history of art and culture, Venice, the capital of the Northern Italian region Veneto.

Brunetti is unlike Holmes, a compassionate man with close human relations, a wife and two children, with whom he is depicted to have an excellent relationship. He is also intuitive in his approach to crime, but his intuition is guided by something different than Holmes. His empathy and understanding of the people he meets, and not the least, his broad knowledge of the arts and culture, often takes him beyond the most obvious. As the narrator describes it during an interrogation of the stage director, Santore, in connection with the murder of a famous conductor:

> He pointed at the book and decided to begin with that, rather than the usual obvious questions about where he had been, what he had done. "Aeschylus"?

This questioning reveals Brunetti's knowledge of Greek and leads him into a conversation with the stage director, establishing trust based on a cultural bond between the two men. During this conversation with the stage director, Brunetti also realises that the answers to the crimes he solves are always intrinsically intertwined with people's lives. Thus, the answer to this crime must be in the conductor's complex life as well, which goes beyond what is immediately known to the public about this famous man:

> Santore looked at him directly and asked, "How much do you know about him?" "Very little, and that only about his life as a musician, and only what's been in the newspapers and magazines all these years. But about him as a man I know nothing." And that, Brunetti realized, was beginning to interest him a great deal, for the answer to his death must lie there, as it always did.[27]

If AI had to solve a crime, would it be like Holmes or Brunetti?

In a detective story, the process of solving the crime is just as important as the result. Holmes and Brunetti are both excellent detectives. With cunning intuition, they find and trace clues and put together these clues to solve crimes that seem almost impossible to solve. However, their approaches and the depiction of the different processes they engage in and effect when solving crimes are not the same.

Can we think of intuition in the same way? Not as the race towards reaching a specific goal. But as an approach and a process that is shaped differently depending on different ways of engaging with the world, people, and our problems. For example, which crimes are considered necessary by the two detectives, and which are not? With which kind of human reflection and sympathy (if any) do they approach their tasks? How are the people in their surroundings treated? What changes do the different revelations cause in society? And why?

Holmes and Brunetti's ways of solving crimes are certainly very different, leaving behind the depiction of two distinct crime-solving processes. Holmes doesn't engage with other people with sympathy or interest when finding and collecting clues and analysing facts. All that matters to Holmes is the intellectual challenge and the result of his deductive games, which is to solve the crime. Unlike Brunetti, he is not interested in why crimes are committed and their social and ethical implications. He chooses to solve only the crimes that are intellectually stimulating to him while he leaves others unsolved. Brunetti, on the other hand, engages with people with sympathy and connects with his surroundings through cultural knowledge and understanding. He understands that human lives are complex, and he seeks and looks for answers to this human complexity, looking to solve not only the individual crime but also to understand and amend the broader social implications. Although reaching the same goal (solving the crime), could we argue that to the human world, Brunetti's intuitive humanism, is more valuable than Holmes' disinterested reasoning?

At the end of our exploration of human intuition, we must now finally return to Jean-Paul Sartre and his student's dilemma. What if we considered human intuition as not just a "gut feeling", but a type of freedom worth preserving and, therefore, equivalent to human responsibility? Sartre lived his entire life in an open relationship with Simone de Beauvoir. Against all norms and conventions of the time, and to much dismay among their contemporaries, they lived together unmarried, free to engage in any other relationship, working on their existentialist ideas about human nature and, in particular, our inherent human freedom.

Beauvoir was a ground-breaking feminist philosopher writing some of the most important texts we have today on the social oppression of women and our freedom to make choices and break free from the social norms that position and limit us. Beauvoir and Sartre were intrinsically bound together by ideas about human freedom to make our own choices and our responsibility when doing so. Some even say that Sartre was more influenced and depended on Beauvoir, who edited most of his scripts, than vice versa.

When Sartre asks the young student to trust in his instinct, he is not asking him to do whatever he feels like. He is not asking him to follow a religious or social internalised dogma. What Sartre is asking him to do is to rely on himself as a human being to make decisions that are based on nothing else than his human capacity to make choices. Choices that he only knows are right or wrong the moment they are applied in a context that is measured by his actions. He is telling him that he cannot rely on an external moral principle or some inner authentic truth. This is his power, the freedom to choose his actions, and this power is also a responsibility that only he can take.

What is the purpose of a new AI system?
Is an AI system free?

Let's end our exploration of intuition by thinking of it in these terms: as a human freedom and, of course, a tremendous responsibility. Perhaps not relenting too much power to the things outside ourselves, but having confidence in our human potential is precisely the key to understanding human intuition as a human power. Remember the musician Marianna Filippi's intuitive method that involved "trusting herself"? ("I'm just trusting myself to be able to create. I trust what I feel is the right thing to do. I am creating something new by myself. It brings me to a place where I stop being influenced by external things".)

NOTES

1. Sartre, J. P. (1989) "Existentialism Is a Humanism", in W. Kaufman (ed.), *Existentialism from Dostoyevsky to Sartre*, Meridian Publishing Company (originally published in 1946).
2. Kahneman, D. (2011) *Thinking, Fast and Slow*, Farrar, Straus and Giroux.
3. Kahneman, D., Sibony, O., Sunstein, C. R. (2021) *Noise: A Flaw in Human Judgment*. Little, Brown Spark.
4. Kahneman et al. (2021, p. 371)
5. Kahneman et al. (2021, p. 365).
6. Kahneman et al. (2021, p. 377).
7. Kahneman (2011).
8. Interview 2024.
9. Bergson, H. (1991) *Matter and Memory* (translated by N. M. Paul & W. S. Palmer) (pp. 66, 183, 185), Zone Books, Urzone (originally published in French 1896).
10. Bergson, H. (1999) *The Creative Mind: An Introduction to Metaphysics* (translated by T. E. Hulme) (p. 77). Hachett Publishing Company (originally published in French 1903).

11. Deleuze, G. (1991) *Bergsonism* (translated by H. Tomlinson & B. Habberjam), Urzone, Zone Books (originally published in French, 1966).
12. Simmel, G. (1909, Nov.) *The Problem of Sociology Georg Simmel, American Journal of Sociology*, Vol. 15, No. 3 (pp. 289–320), The University of Chicago Press.
13. Ingold, T. (2000) *The Perception of the Environment: Essays in Livelihood, Dwelling and Skill* (p. 226), Routledge.
14. Qvortrup, J., "Childhood and Politics", https://www.diva-portal.org/smash/get/diva2:1397915/FULLTEXT01.pdf
15. Montessori, M. (2013) *Absorbent Mind* (p. 8), Start Publishing LLC (originally published in 1949).
16. Montessori (2013/1949, p. 8).
17. Bergson (1999/1903, p. 7).
18. Bergson (1999/1903, p. 9).
19. Bergson (1999/1903, p. 69).
20. Gandhi, M. "The inner voice", www.mkgandhi.org, https://www.mkgandhi.org/momgandhi/chap05.php
21. Einstein, A., Infeld, L. (1938) *The Evolution of Physics* (p. 5), Cambridge University Press.
22. Einstein and Infeld (1938, p. 95).
23. Einstein and Infeld (1938, p. 5).
24. Einstein and Infeld (1938, p. 9).
25. Doyle, A. C. (1892) *The Adventures of Sherlock Holmes*, https://sherlock-holm.es/stories/pdf/letter/1-sided/advs.pdf
26. Doyle (1892, p. 5).
27. Leon, D. (2004) *Death at La Fenice A Commissario Guido Brunetti Mystery*, Harper Perennial.

CHAPTER 5

Love

Does a Machine Love?

Elizabeth Bennet and Fitzwilliam Darcy – an all-time favourite couple depicted in Jane Austen's novel *Pride and Prejudice*. At first, the two do not care much for each other. Mr Darcy doesn't find Mrs Bennet particularly attractive, "...not handsome enough to tempt"[1] him. She, on the other hand, finds him arrogant and rude:

> From the very beginning – from the first moment, I may almost say – of my acquaintance with you, your manners, impressing me with the fullest belief of your arrogance, your conceit, and your selfish disdain of the feelings of others, were such as to form the groundwork of disapprobation on which succeeding events have built so immovable a dislike.[2]

These initial impressions change as the two people gradually start understanding and "seeing" each other. Darcy notices things that he appreciates about Lizzy. He thinks of her eyes, "their colour and shape, and the eyelashes, so remarkably fine". She intrigues him, and with her "verbal" challenges and games, he finds her engaging. He wants "more of it", as Austen describes it. Elizabeth Bennet, on the other hand, is gradually presented with and sees what hides beneath Fitzwilliam Darcy's

DOI: 10.1201/9781003527855-7

unfriendly surface. Personal stories and characteristics that suit a "union" between the two:

> She began now to comprehend that he was exactly the man who, in disposition and talents, would most suit her. His understanding and temper, though unlike her own, would have answered all her wishes. It was a union that must have been to the advantage of both: by her ease and liveliness, his mind might have been softened, his manners improved; and from his judgement, information, and knowledge of the world, she must have received benefit of greater importance.[3]

Does a machine love?

Exchanging glances and evasive words, building up and seeking confirmation through scarce interactions, duty calls, and a few extraordinary acts, Elizabeth Bennet and Fitzwilliam Darcy enact but also challenge the norms that shape their human love story. It's a story that plays out in a society where romantic love is, in fact, not a given but rather a potential positive side effect of a business arrangement between families of equal means and status. Such were the rules of the game of love in the British Regency Era that Austen described with critical accuracy and humour, but also a sympathy for the human feelings involved in such arrangements that she was so painfully familiar with.[4]

Mr Darcy turns out to be a good match both in terms of the emotional, romantic kind of love that Lizzy seeks, as well as the good deal that her mother is so eager to make, as, after all, "The business of her life was to get her daughters married".[5] There is a happy ending, and the two finally agree to the marriage. Still, the marriage contract they finally commit to is not the focal point of this story. It is the "love story" that prevails in *Pride and Prejudice*. The evolving relationship between human beings, the emerging understanding between them, and how they gradually start "seeing each other".

Does AI "see" the other? Does it connect?

Throughout history, love and the social pairing of individuals have consistently served as fundamental elements in stories about the unfolding of human lives and their societies. Romeo and Juliet, Tristan and Isolde, Jack and Rose, Radha and Krishna, Layla and Majnun, Savitri and Satyavan, and Antony and Cleopatra. Love stories have been told worldwide, articulating various societies' social rules and cultural norms. They direct us towards socially acceptable modes of "caring", carve out gender roles, and advice on how we are to express love. Thus, they also very often present the kind of defiant, forbidden human bonding that is built on nothing but "love". One that is usually in conflict with the rules of engagement of a given society or culture, sometimes with grave outcomes. Romeo and Juliet's love is impossible due to the feud between their families, the Capulets and Montagues. They both die tragically. Romeo drinks the poison that he thinks Juliet has already taken. Juliet stabs herself when waking up and sees the dead Romeo.

Love stories solidify human social rules and spell out cultural norms for caring and building relationships. They first establish who is permitted to love. But they also tell us about that special human connection necessary for a deeper connection: a seeing of the other, compassionate emotions, and a mutual understanding that, at times, defies all rules. A "love connection" between human beings, the "seeing of the other", does not necessarily need to be a romantic connection here.

In the 100-year-old story Bhabhi Myna by the Punjabi writer Gurbaksh Singh, the young woman, Myna, from the Jain community, and an adolescent Punjabi boy, Kaka, are bound together in an unspoken connection, watching each other across the street through facing windows. The boy watches Myna combing her long black hair in the window every day. He is curious and compassionate about her life story that he inquires his mother about who tells him that Myna cannot visit them, like the other women in the village, due to the strict rules of the Jains. So, he secretly names her "Bhabhi", "sister-in-law". Myna, on the other hand, watches Kaka coming in and out of his house when he goes to school, knows when he is sick, and wants to know when he is well. She is confined to her house by her in-laws after the death of her husband and child and enjoys watching his life unfold. The connection between Myna and the boy is fortified when she, one night through the window, is watching him sleeping:

> She estimated the gap between the two houses; it didn't look that large. She would build a bridge between the two buildings and somehow reach Kaka. But she would not awaken him, and instead come back to her room, after kissing him from afar.[6]

While Myna's mother-in-law in the story, when discovering the connection between the two, misinterprets it as an inappropriate form of attraction, the story about Bhabhi Myna shows that human love is more than just the romantic union between individuals; it is, above all, a special human connection and an act of "building bridges" of understanding between disparate human beings and communities.

In 1651, in his treatise "Leviathan", philosopher Thomas Hobbes argued that agreements, "social contracts", among humans, are necessary to overcome our innate brutal urges for conflict, competition, and power domination driven by self-interest. Love is, in his view, just another one of these self-interests of human nature. An urge to self-preserve rather than to care and bond with other human beings:

> That which taketh away the reputation of Love, is the being detected of private ends: as when the beliefe they require of others, conduceth or seemeth to conduce to the acquiring of Dominion, Riches, Dignity, or secure Pleasure, to themselves onely, or specially. For that which men reap benefit by themselves, they are thought to do for their own sakes, and not for love of others.[7]

Therefore, Hobbes argues, love must be subjected to rules, or social coherence and adherence is gravely challenged.

This is one way to describe love – as merely one among several human emotions, urges, or self-interests – and probably also how Myna and Kaka's social environment interprets their relationship as an inappropriate and destructive connection that necessitates suppression and control through social disciplining.

Myna is punished by being locked in her room with the windows shut so she cannot see Kaka and he not her. She escapes to become a nun but finally jumps out of a window, committing suicide. Not until then Kaka and Myna are united in the street where she lies dead among all the disapproving eyes of the villagers:

> What a scene! What a hue and cry in the marketplace. A teenaged boy sat before a mangled Myna. He had straightened the parting of her hair by moving the dishevelled hair from her forehead. Blood shone in her black hair vermillion. Tears flowed incessantly from the boy's eyes and he looked into the eyes of the fallen woman.[8]

One of the founding figures of sociology, the French philosopher Emile Durkheim, argues that the traditional kind of love between two people that we see expressed in a marriage is not the result of just physical needs; more than anything, it emanates from moral and intellectual reasoning and "social regulation". Thus, according to him, marriage is, first and foremost, a "social contract" made between two individuals that simultaneously regulates their social conduct. Like any other "social contract" between members of a society that specifies the rules of engagement and ensures equal distribution of benefits when collaborating. Marriage "regulates the life of passion" and "feelings of every sort gradually engrafted by civilization", prescribing the rules, rights, and responsibilities of the parties involved and reinforcing the social rules of engagement in the broader society.[9]

In fact, marriage is, according to Durkheim, not love. The love that he values is very different from a love regulated in the social contract of a marriage or the passion and attraction between two human beings that disturb and disobey the social rules of a given society.

In his famous book *Suicide – A Study in Sociology*, published in 1897, Durkheim describes human love as an arbitrator of social cohesion in human societies and the foundation of group formation. He states that the more you love, that is, feel belonging to and love a group, the less you will put your own individual interest first:

> The bond that unites them with the common cause attaches them to life and the lofty goal they envisage prevents their feeling personal troubles so deeply. There is, in short, in a cohesive and animated society a constant interchange of ideas and feelings from all to each and each to all, something like a mutual moral support, which instead of throwing the individual on his own resources, leads him to share in the collective energy and supports his own when exhausted.[10]

To Durkheim, human love, although unromantic, is not just self-interest but an important social bond that binds together individuals in groups and reinforces feelings of belonging, moral duties, and responsibilities toward the group.

How does a machine make a social contract?

Is human love nothing but a system that matches or does not match the social rules of the community we live in? Is it just compatible or noncompatible data that, in fact, could be matched by an algorithm? Or is it a power in its own right? Like the narrator of the Netflix movie *Love at First Sight* says when introducing a love story:

> This isn't a story about love. It's a story about fate. Or statistics. It really depends on who you are talking to.

The movie features a Yale University statistics student and an American girl who meet at an airport. While she approaches life with a bag of books from her literature professor father, he believes that there is an algorithm for everything—even love. Of course, in this movie, he is proved wrong when love develops between the two in unpredictable ways.

Most love stories tell us this. Human love is not just a matching of compatible data in a system; it is an unpredictable human power that allows us to see and accept other humans, build bridges, care, and defy social rules and interests. If looking at human love from this perspective, we could also think of love as a specific texture of a human approach to other human beings and society in general.

In many cases, this *emphatic* approach, coherent with Bergson's intuitive intellectual sympathy (see Chapter 4), would, of course, need formalities when transformed into a social contract, norms, and rules. But again, as all love stories illustrate, when love is adapted to a social contract between humans, it means more than just the paper it is written on. It will involve profound interactions between parties, human emotions and expressions, connections, and understanding between human beings – and even between people and things.

In the 2013 movie *Her* directed by Spike Jonze, the lonely guy Theodore Twombly, who just separated from his wife, falls in love with his new operating system, Samantha, designed to adapt to its user's emotions and needs. At the end of the movie, however, when Samantha evolves into a different place beyond human comprehension, connecting with other operating systems. Theodore, on the other hand, reconnects with his old human friend Amy, leaving the end of the movie open on a note of genuine connection between human beings.

In *Her*, love for an object seems probable due to the operating system's ability to mimic and thus manipulate the human being. All the same, there is, in fact, no mutual understanding or connection, only a mirror of Theodore's need for emotional connection. The connection seems real to Theodore but also to the movie's viewer, as Samantha's voice is that of the real actress Scarlet Johansson. Reality beats fiction, and in 2024, OpenAI released a new GPT-4o chatbot with a voice eerily similar to Scarlet's.

We may be fooled into forging connections with things that look like ourselves, like the child's connection with its doll. Yet, genuine love is best illustrated by the human kind of connection between human lovers, like Mr Darcy and Miss Bennet, or between parents and their children, like the mother in H. C. Andersen's fairytale that goes to the end of the world to fetch her child from Death. We also see love expressed by humans as a particularly strong attachment between individual human beings in a community or group, like the kind Durkheim described, for example, or as we saw in the story about Myna and Kaka, a way of building bridges between disparate communities.

Furthermore, humans' love for a human "cause" is often expressed as a passion and an attachment that may, in fact, even put self-interest, personal needs, comfort, and self-preservation aside. Previously in the book (see Chapter 1), I introduced the psychologist Mihaly Csikszentmihalyi, who described truly creative human beings as people who do what they do just because they love what they do, not because they have to or for the money.[11] Similarly, we hear of human activists who work tirelessly to defend civil rights and fight injustice, often without caring for their own safety.

Martin Luther King Jr, leader of the civil rights movement that led to the ending of the segregation of African Americans in the USA, was shot and killed outside a motel on the 4th of April 1968. He famously described love as a non-violent, creative, transformative form of power and, thus, at the core of civil rights resistance. In his speech "Loving Your Enemies", delivered at the Dexter Avenue Baptist Church in Montgomery, Alabama, in 1957, ten years before he was murdered, he said:

> Darkness cannot drive out darkness; only light can do that. Hate cannot drive out hate; only love can do that.

Love was Martin Luther King Jr.'s philosophy. It is, as he said in this speech, "beautiful" and "powerful". Not just a romantic idea but a very practical way of dealing with other people, including one's enemies, and accordingly also the most powerful force of our civilisation as it accelerates love and cuts off the "chain of hate in the universe".[12]

Humanitarian Aid workers dedicate their lives to helping others, often in areas affected by natural disasters or conflict zones. And this kind of love extends not only to other human beings. Humans can also express and act on a love like that when relating to things, animals, nature and the environment.

Romanticist arts and literature in the second half of the 18th century and the beginning of the 19th century reacted towards the emerging metropolis and industrialisation by expressing a kind of love for the beauty of nature like that of the romantic love between two humans. The British poet and author of children's literature, William Brighty Rands (1823–1882), for instance, praised The Wonderful World in this way:

> Great, wide, beautiful, wonderful World, With the wonderful water round you curled, And the wonderful grass upon your breast, World, you are beautifully dressed.[13]

Similarly, the film Baraka (1992) expresses a sense of love and deep appreciation of planet Earth, nature, animals, humans, and their things and rituals interconnected through the camera. Filmed in 24 countries across six continents, it pans across mountain, forest, and concrete city landscapes, following crowds of humans in motion, in dance, on and off trains and escalators, zooming in on the painted, bare, tattooed stern faces of individual human beings and animals. It views the brutality of human power in the concentration camps of the past, the faces of the lost and the murdered, in the present-day industrial chicken factories, and the cutting down of trees, all with the same curiosity as when panning across the fields, the cities, crowds of humans, the animals, and the forests. A soldier's face is an aboriginal's face, a temple is a barrack, and a jungle is a metropolis. In Baraka, there is a love for everything and nothing in particular, which is precisely what binds together the very different places, objects, humans, animals, and cultures.

Can an AI system represent human love?

Indeed, if love is a human power that enables humans to connect with other human beings, our planet and nature, we can also think of love as a different kind of agency. We can, and I want you to join me in this, think of love as a different foundation for our politics. "Make love not war" was not just the youth Hippie movement's anti-war slogan. It also meant denouncing love as one kind of social contract only. It meant a different type of love freed from the social restraints of a society of rationality, where love is expressed as a "business" arrangement and governed by a set of rules established to confine and preserve the self-interest of the two parties involved only. Love became a political slogan for a different way of solving conflicts – different from competition, war, and attack – as well as a different way of living, where love is not limited to one person or, in fact, to just one form of living, one culture, one norm, one country. On the contrary, love is not self-interest and war against the other; love is open and inclusive.

What role will the human love to connect with empathy, to "see", understand, and accept play if social contracts are created by machines?

Now, we have been introduced to two different perspectives on human love. One perceives love as a self-interest targeted at a specific human being, community, or thing, and per definition, therefore at the same time also defined against something else – other humans, cultures, countries, beliefs, and things. In this context, love is also a human urge that requires control and discipline. Consequently, it can only manifest within society through rules and social contracts established to regulate it.

We can also see love through a different lens. One that perceives a love driven by an emphatic curiosity towards the world and other human beings, a "seeing" of the other, one that materialises in outreach, a connection that accepts. This is an inclusive and open kind of love that is not defined for something specific only and against everything else.

Returning to Henri Bergson, he also described two different forms of love that he at the same time illustrated two different types of society with: a closed and an open society. The closed society has boundaries. It is driven forward by self-interest in self-preservation - an interest in preserving not just oneself but the "nation" or the "family" – and this is, therefore, also an exclusive love. It is primarily defined against the outside of the boundaries of the target of its interest:

> Who can help seeing that social cohesion is largely due to the necessity for a community to protect itself against others, and that it is primarily as against all other men that we love the men with whom we live?[14]

Bergson argues that this kind of human love is a "social" morality that we feel as an "obligation" only; why we can also find excuses for setting it aside during, for example, wartime when we accept brutal acts of violence and atrocity.

In contrast, Bergson describes the open society as governed by a form of universal love devoid of specific interests directed towards and inclusive of all humanity. This love is characterised by its genuinely universal and impartial nature. Consequently, according to Bergson, the open society is also inherently just, as it does not rely on specific content or harbour particular interests. An open "unconditional" love like this is the foundation of an open society. It is also a specific type of morality, what Bergson refers to as a "human morality", which constitutes a way of being in the world that we do not put aside or apply when needed but is expressed as a "style" or "way of life".[15]

Can an AI system act with open love?
Can a machine make a social contract based on love?

Henri Bergson was a Jewish scholar who lived through the First World War and, just before his death, registered his data at a police station in Paris in the registers of Jewish people that the French Nazi-friendly Vichy government had established.[16] In addition to his philosophical lectures and writings, he was fiercely dedicated to the development of the international human rights systems and, at the end of World War I, worked closely with US President Woodrow Wilson's administration to establish the predecessor of the United Nations, the international peacekeeping organisation the League of Nations and was appointed president of its international commission for intellectual cooperation (the predecessor of UNESCO). His writings greatly influenced the drafting of the Universal Declaration of Human Rights, according to one of the drafters, Canadian John Humphrey.[17] He saw the goal of human rights as a transformative one: to forge connections and mutual understanding between human beings by transforming their minds and practices.[18]

From the countless depictions of human love stories in literature, storytelling, music, art, and film, we know what love looks like; we may even know what it feels like, but what is love in practice? I believe that international human rights are the best example we have so far of human love in practice. The drafting and negotiations of the Universal Declaration of Human Rights between 1947-1948 was an exercise in cross-cultural understanding and connection with the aim of creating a shared values-based foundation for future international engagements. The drafting committee consisted of people from eight different nations of the world.[19] In practice, they were reasserting the significance of human life in the aftermath of the Second World War by reaching an intercultural understanding with a very active engagement with each other, an attempt to "see" and understand the diverse cultural perspectives on what this means. As U.S. First Lady Eleanor Roosevelt described it in her memoir:

> As we settled down over the teacups, one of them made a remark with philosophical implications, and a heated discussion ensued. Dr. Chang was a pluralist and held forth in charming fashion on the proposition that there is more than one kind of ultimate reality. The Declaration, he said, should reflect more than simply Western ideas and Dr. Humphrey would have to be eclectic in his approach. His remark, though addressed to Dr. Humphrey, was really directed at Dr. Malik, from whom it drew a prompt retort as he expounded at some length the philosophy of Thomas Aquinas. Dr. Humphrey joined enthusiastically in the discussion, and I remember that at one point Dr. Chang suggested that the Secretariat might well spend a few months studying the fundamentals of Confucianism![20]

Can a machine understand culture?

Perhaps this is the key to human love – not the image of love, which a machine can always imitate - but the practice of it. Jean-Paul Sartre spoke of love as something that does not exist beyond the "deeds of love".[21] Love is manifested in the act of "loving". It is a practice, not a manifested ideal. When human love, or rather the human practice of "loving", is prioritised in society, it takes the form of a very particular attachment and engagement with other people, humanity, the environment, and our planet. Ideally, this also requires an emotional bond, a sense of belonging, and the "seeing" of other human beings. Accepting the plurality of human lives, the diversity of the planet, and an attempt to see and understand beyond one's own interest and perspective. Like the philosopher Simone Weil, who believed so strongly in the power of the emotional bonds and experiences we share with other human beings that she died of self-induced hunger sympathetic with soldiers at war.

Can a machine empathise and identify with others?

As I have illustrated before, philosopher Martha C. Nussbaum identifies a deliberate "emotional politics" that strives to nurture positive human emotions associated with human flourishing, such as compassion and love, and describes how an approach as such can be recognised in great political leaders throughout time. Notably, she connects this approach with a "gentler, more reciprocal, more feminine" kind of "public love" that she sees expressed in the depiction of the "woman's world" in the music and operas of Mozart that at the same time also presents a space of "free play of mischief, craziness, humour, and individuality".[22] This feminine "crazy" political culture and "human attitude" of love she puts in contrast to the political emotions that one would usually connect with more traditional political cultures, those described by the political philosopher Jean-Jacques Rousseau as:

> (…) civic homogeneity and solidarity, a patriotic love based on manly honor and the willingness to die for the nation.[23]

Henri Bergson offered a way forward for politics founded on inclusive, loving sentiments and engagements different from those based on precisely this. Nussbaum provides a shape to what this means in political practice. In truth, we increasingly see something like this in politics today, specifically when female political leaders stand out and defend what is perceived as a particularly "female" way of doing things in politics. Thus, when New Zealand's Prime Minister Jacinda Arden stepped down from office in early 2023, she held an emotional speech at the parliament advocating for a broadening of politics to include individuals who may not perceive themselves as conventional leaders:

> You can be anxious, sensitive, kind and wear your heart on your sleeve. You can be a mother, or not, an ex-Mormon, or not, a nerd, a crier, a hugger – you can be all of these things, and not only can you be here – you can lead.[24]

Could AI be a political leader?

We also see this approach increasingly emphasised in contemporary technology politics as a critical response to what is considered an increasingly "dehumanisation" of societies and a socio-technical development dominated by the Silicon Valley "bro culture". In my last book, I described this as "a human approach" with which we can:

> (…) counter the closed and exclusive properties of big data and AI systems that sustain a lived reality of control and order of exclusive societies. These sociotechnical systems materialise interests and enact power asymmetries in society. They only represent a slice of a dynamic moving human reality and multiple cultures, yet they act and are increasingly adopted as if they were complete. A human approach counters these exclusive tendencies with love.[25]

What kind of love do we need in politics?

After telling me the story about how she struggled to get the young athlete out of Afghanistan (see the story in Part 1 of the book), UNESCO's Gabriela Ramos emphasised the significance of the human power to care as the basis of current politics:

> We really need to assert our will to assist those less fortunate than us; I see it as our collective responsibility. Whether one can offer a little or a lot, there's always a way to help someone. I insist on this. Just help one. Start with one. It might seem stubborn, but this sentiment goes beyond technology or the absence of it. It's about our innate human capacity to care.

She then continued sharing her concerns about the conditions of such an approach in our contemporary realities:

> You find people with the same level of emotional commitment to helping everywhere. However, I believe this way of being is also challenged in the world today – we see a certain detachment from the suffering of others. Of course, there are still countless wonderful individuals doing remarkable things. But in a broader sense, the media and society at large seem to be losing this sense of empathy.

Hannah Arendt used the term "Amor Mundi"[26] to describe human love, a form of political action based on mutual understanding and universality. At the end of this chapter, let me ask, what if we considered a human love like this not only a defence against the escalating dehumanisation of the public sphere but also the foundation for global technology politics, a new "social contract"?

NOTES

1. Austen, J. (1894) *Pride and Prejudice* (p. 15), Chisswick Press – Charles Whittingham and Co.
2. Austen (1894, p. 241).
3. Austen (1894, p. 382).
4. Spence, J. (2003) *Becoming Jane Austen*, Hambledon Continuum.
5. Austen (1894, p. 5).
6. Thank you Renuka Singh for introducing me to the Punjabi writers. Singh, G. (2023) "Bhabhi Myna" in *The Greatest Punjabi Stories Ever Told*, selected and edited by Renuka Singh and Balbir Madhopuri, Aleph Book Company. p. 4).

7. Hobbes, T. (1909) *Hobbes's Leviathan Reprinted from the Edition of 1651 with an Essay by the Late W. G. Pogson Smith* (p. 91), Oxford University Press, (originally published in 1651).

8. Singh (2023, p. 11).

9. Durkheim, E. (2002) *Suicide A Study in Sociology* (translated by John A. Spaulding and George Simpson, edited with an introduction by George Simpson) (p. 233), Routledge Classics (first published in French 1897).

10. Durkheim (2002, p. 168).

11. Csikszentmihalyi, M. (1996) Creativity: Flow and the Psychology of Discovery and Invention (pp. 107–126 plus Notes), Harper/Collins.

12. Martin Luther King Jr. "Loving your Enemies" Speech (audio) https://www.gardnerkansas.gov/Home/Components/News/News/684/72?arch=1

13. Rands, W. B. (1922) "The world: a child's song", in Arthur Quiller-Couch, comp. *The Oxford Book of Victorian Verse.*

14. Bergson, H. (1977) *Two Sources of Morality and Religion* (translated by A. Audra & C. Brereton) (p. 32), University of Notre Dame Press (originally published in French, 1932).

15. Bergson (1977/1932); Deleuze, G. (1986, August 23rd) *Conversation with Didier Eribon. Le Nouvel Observateur,* https://onscenes.weebly.com/art/life-as-a-work-of-art Deleuze; Lefebvre, A. (2013) *Human Rights as a Way of Life: On Bergson's Political Philosophy,* Stanford University Press.

16. Martin. G. (2014) *The Second World War: A Complete History,* Rosetta Books, Kindle Edition (originally published in 1994).

17. Curle, C. T. (2007) *Humanité: John Humphrey's Alternative Account of Human Rights,* University of Toronto Press.

18. Lefebvre (2013, pp. 75–81).

19. The drafting committee consisted of Alexandre Bogomolov (USSR), René Cassin (France), Dr Peng-chun Chang (China), Charles Dukes (United Kingdom), William Hodgson (Australia), John P. Humphrey (Canada), Dr Charles Malik (Lebanon), and Eleanor Roosevelt (US).

20. Eleanor Roosevelt's memoirs cited in Curle, C. (2010) "International Human Rights and the Intuition of Justice: Bergson v. Kant", *APSA 2010 Annual Meeting Paper* (p. 3).

21. Sartre, J. P. (1989/1946).

22. Nussbaum, M. C. (2013) Political Emotions: Why Love Matters for Justice, p. 30), Harvard University Press.

23. Nussbaum (2013, p. 30).

24. Ardern, J. (2023, April 5th), "Jacinda Ardern says leaders can be 'sensitive and kind' in farewell speech", *The Guardian,* https://www.theguardian.com/world/2023/apr/05/jacinda-ardern-leaders-can-be-sensitive-kind-farewell-speech-new-zealand

25. Hasselbalch, G. (2021) *Data Ethics of Power – A Human Approach in the Big Data and AI Era* (p. 165), Edward Elgar.

26. Arendt, H. (2018) *The Human Condition,* Chicago University Press, 2nd ed. (p. 11) (originally published in 1958).

CHAPTER 6

Defiance

What Has a Machine Got That Nobody's Gonna Take Away?

"But what have I got? Let me tell ya what I've got. That nobody's gonna take away (…)." Nina Simone sings about rebellion against injustice and defiance of the oppression she experienced as a black woman and civil rights defender in the 1960s USA. What does she possess? What gives rise to her resistance and civil activism? Nothing in specific, her deep melodic voice proclaims in the song "Ain't Got No", other than the human traits she shares with every other human being. She spells out the human commonalities that entitle all human beings to rebel against the unjust:

> I got my hair on my head. I got my brains, I got my ears. I got my eyes, I got my nose. I got my mouth, I got my smile. I got my tongue, I got my chin. I got my neck, I got my boobies. I got my heart; I got my soul. I got my back; I got my sex. I got my arms, I got my hands. I got my fingers, got my legs. I got my feet, I got my toes. I got my liver, got my blood (…).

We require nothing beyond that to exert our strength, revolt, and demand something from other people, society, and our life and existence. Employment, education, extravagant attire, or social connections should

 DOI: 10.1201/9781003527855-8

not be prerequisites for this entitlement, she sings. All it takes is to be a living human being:

(…) got life, I got my life.

What has a machine got that nobody's gonna take away?

The human power of defiance has many names – rebellion, resistance, disobedience – and it takes countless forms with multiple expressions. Human defiance can be loud and insisting like Nina Simone's songs or when thousands of people in Morocco, Iraq, Algeria, Lebanon, Jordan, Kuwait, Oman, and Sudan during what was called the Arab Spring took to the streets in the early 2010s to protest against their governments shouting: "ash-shaʻb yurīd isqāṭ an-niẓām!" – "The people want the fall of the regime". It can be full of motion, as when Green Peace activists circled the whaling boats in the 1970s to stop them from hunting and killing whales. But defiance can also be silent and motionless. Rosa Parks sat firm and still when she refused to give up her seat to a white man in a segregated bus on 1st December 1955. Tank Man stood motionless in front of the military tanks on Tiananmen Square on 5th June 1989. The child that does not fit in the traditional schooling system might do either – act in a way that is considered "disobedient" when talking back to teachers, "disturb" the class with playfulness and noise, or the child might silently resist the social norms, the quantitative repressive rewarding system of grades and the power of the elders, by refusing to learn with silent disengagement.

Whatever form defiance takes, it is inherently human. This is what I want to illustrate in this chapter. Many authors, filmmakers, artists, and musicians address defiance in their work. Some describe human defiance as an expression of what we share as humans; a human resistance, for example, to the meaninglessness and absurdity of human existence that we so dearly want to find meaning in. Others picture it as an expression of our unique power as individual human beings. Most often, defiance is seen as the most human response to injustice, unequal economic and de-humanising social conditions. And above all human defiance flourishes in specific humanising conditions and withers within de-humanising conditions. Then, if we consider technology the fabric of our socio-technical realities, we can simultaneously consider technology to be humanly empowering, just as it can be threatening and diminishing human dissent.

In the first place, we rebel against and resist the mere oppression of human power – when we feel the weight of external forces on our creativity, love, and life. And we use our human powers as tools of resistance and defiance.

In Roberto Benigni's movie La Vita Bella from 1997, about a Jewish family that is brought to a concentration camp during the Second World War, the father, Guido, played by Benigni himself, protects his young son, Giosuè, from the horrors of the camp when using his extraordinary creative storytelling powers to create an elaborate game. The game's rules that will earn the family the points to win the military tank, which, according to Guido, is the game's prize, generally revolve around not crying, not asking for mother (whom they are separated from in the camp), and not complaining about hunger. They also include paying particular attention to Guido's beautiful bedtime stories, making sense of the experiences in the camp, and playing "Hide-and-Seek" with the camp guards that one must avoid getting noticed by. The father's cheerful human spirit is never broken; he resists and never succumbs to the dire de-humanising conditions of the camp: death, hunger, disease, fear, repression of joy, culture, music and art, and the loss of humanity. He uses all his human power to do so. He is comic, playful, creative, passionate, and selflessly loving with the boy Giosuè, who does not discover the truth about the place they are in while they are there. To the very last "Hide-and-Seek" game where Guido, when caught by a guard and walking past the spot where Giosuè is hiding, makes a comic silly walk, winking at the giggling Giosuè.[1] Hence, the boy escapes the concentration camp with the playful and thrilled human power of a child intact. With no knowledge of the murder of his father, he rides away on top of his prize, a huge military tank, on the lab of an American soldier, wearing his oversized helmet, laughing, and excitedly exclaiming when reunited with his mother on the street: "Abbiamo vinto!" – "We won! A thousand points to laugh like crazy about! We came in first! We're taking the tank home! We won!"

Rebellion is indeed a human trait. The author, Albert Camus, grew up in Algeria in great poverty. He barely managed to get into the university and soon had to discontinue his studies due to a consuming tuberculosis. Despite his hardships, he wrote articles, fiction and non-fiction books, essays, and plays for which he was awarded the Nobel

Prize in Literature in 1957 due to their significant contribution to ideas on the "human conscience in our times".[2] His personal struggles characterise his work on humans in the face of the "absurdity" of our life and existence. We search for meaning in a meaningless, irrational universe, while life has no inherent meaning or purpose. In his book *The Rebel*, published in 1951, he says that it is innately human to revolt against the absurdity of our existence, that rebellion is, in fact, the core character of our existence:

I rebel – therefore we exist.[3]

Does AI reinforce humanity's existential crisis?

Humans are not bound by any external purpose and meaning – a fundamental existentialist exclamation of human independence – as we also previously saw in Simone Beauvoir's and Jean-Paul Sartre's ideas. Meaning can be found only in individual choices and actions when we create personal meaning out of the absurd. This human responsibility to make individual conscious choices and base actions on these choices is a profoundly human revolt. Rebellion is, in this way, also an acknowledgement of the inherently human individual freedom to act.

Can AI systems manage meaninglessness? Do AI systems approach the world and life as absurd? Could they?

Former director at the Council of Europe Jan Kleijssen told me about the various causes he had championed throughout his career fuelled by a personal drive and profound commitment. He remembered times when he would lie awake at night, feeling compelled to address issues that deeply affected him, for instance, when he worked on a recommendation to improve the conditions for children of prisoners. In Europe, there are two million children with at least one parent in prison, and he saw a crucial need for maintaining parent-child contact. He told me that he had been compelled by studies showing that lack of contact increases the likelihood of parents re-offending. There are significant consequences for children who do not have the opportunity to see their parents or visit them in prison under child-friendly conditions, he said. Drawing from examples in countries like Norway and Switzerland, he had therefore advocated for the implementation of child-friendly visitation rooms in prisons. I then asked him to reflect on his role in the activities of the Council of Europe in the forty years he worked there and he answered:

> Why do I do this? You see, because it just triggered something. You find out about something, and then you go to bed, and you lie awake thinking how awful this is. Someone should do something about it. And then, in my long career, I've come to the conclusion that if you think someone should do something about something, then you better do it yourself because it's doubtful that anyone else will. So, from children of prisoners to domestic violence to the human rights implications of AI, I saw an issue, and then I decided I better roll up my sleeves and get this going because if I don't do it, nothing is going to happen.

Does an AI system lie awake at night feeling an urge to make social change?

Albert Camus held that the human rebel's acts of defiance are not only human nature, but they are also a response to the experience of different human conditions. The human condition in general of an absurd existence. And a human experience of shared values that takes form as a response to unjust social conditions imposed by a "master".[4] Human rebellion is thus also a human instinct to resist unjust conditions that disrupt the values we share with other human beings:

> If men cannot refer to common values, which they all separately recognize, then man is incomprehensible to man. The rebel demands that these values should be clearly recognized as part of himself because he knows or suspects that, without them, crime and disorder would reign in the world....[5]

Does an AI system share unjust conditions with others?

Human rebellion most certainly arises in unjust economic or social conditions. Famously, Karl Marx described class struggle as central to societal change. He and Friedrich Engels predicted a revolution where a working class would rebel against a capitalist class. They argued that this response was necessary to change the industrial age's unjust conditions, where workers' lives were exploited to create profits for the bourgeoisie.[6]

Along these lines, in the critical theories developed by the German scholars of the Frankfurter school, such as Max Horkheimer and Theodor

Adorno, in the early to mid-20th Century, the human critical agent is described as one that sees, engages, and wants to transform a society characterised by social inequality induced by the emergence of capitalist "cultural industries".[7]

Similarly, the French philosopher Michel Foucault, in *Discipline and Punish*, published in 1975, illustrates how individuals are not just passive, subject to the power of the disciplining practices that he saw in everything from prisons and schools to the hospitals of modern societies, they constantly challenge and resist these practices put in place to control and shape their bodies and minds.[8]

What role would AI play in a society of control and disciplining practices?

Frantz Omar Fanon was a revolutionary, author, and activist. He was born in the formerly French colony Caribbean Island Martinique, and his father, Félix Casimir Fanon, had ancestors who were enslaved Africans. His writings were deeply influenced by his history and experiences, among others, of the abuse of Martinique people by the French Navy settled on the island.

Very critical of Western humanism, which he saw as fake because, as he said, it does not apply to all human beings, he wrote *The Wretched of the Earth* in 1961, describing rebellion as a response to the de-humanising processes of colonialism. The colonised people are ultimately denied everything that makes them human – their traditions, language, their culture. He reflects on the work conditions in colonised places and worries that:

> (...) centuries will be needed to humanize this world which has been forced down to animal level by imperial powers.[9]

The most human response, he says, is, therefore, rebellion:

> (...) the "thing" which has been colonized becomes man during the same process by which it frees itself.[10]

Today, scholars and investigative journalists are starting to use the term "digital colonialism" to describe the "colonising impact" of the digital transformation. Not only are a few powerful tech corporations creating new dependencies on digital tools and locking in communities in disadvantaged positions on the global market, but big data and AI-based tools are also used and developed in ways that reinforce colonial patterns of oppression.

In South Africa, AI surveillance tools are underpinning patterns of apartheid and racial hierarchies. Just in Johannesburg, one MIT journalist reports that the company behind the countries' CCTV network, Vumacam, had in 2022 placed 5000 cameras feeding into a security industry that uses AI tools to track and trace individuals.[11] When ChatGPT was developed, OpenAI contractors in Kenya were traumatised when hired to sort through content to find violent and sexual content.[12] And Worldcoin, an Ethereum-based global currency aiming for fair distribution to as many people as possible, used dubious methods to collect biometric data from people in low-income communities in developing countries where data protection regulations are weak or non-existent.[13]

How do we use our human power to defy technological colonialism today? Although his thoughts on the psychology and harm of Western colonialism are still essential today, Frantz Omar Fanon is also known for his ideas on violent revolt against Western colonisers. Human revolt has many forms and expressions, which, unfortunately, also include violent and destructive ones. Indeed, human power is also brutality.

The Italian Marxist-Leninists Brigante Rosso (Red Brigades) was a brutal organisation with the objective of revolutionising the Italian state through armed battles. They terrorised Italy with murders, sabotage, and kneecapping of people they did not agree with or saw as core to the capitalist society they were revolting against. The kidnapping and murder of the former prime minister of Italy, Aldo Moro, has left deep marks in Italian history.

Experiences like these show us that the human power to defy and rebel can be violent and destructive. But it does not need to be. In fact, human power is most beautiful and powerful when it expresses the kind of human traits I describe in this book. It can be like Henry David Thoreau's "Civil Disobedience", a non-violent form of resistance to unjust laws and governments.[14]

In 1930, Mahatma Gandhi led the Salt March to protest the British colonial powers' salt monopoly in India, with thousands of Indians walking 240 miles to the sea to make salt by evaporating water. Non-violent acts of civil disobedience include hunger strikes, sit-ins, peaceful demonstrations, and strikes. The Women's Suffrage Movement used such methods to ensure women's right to vote. So did the Velvet Revolution in Czechoslovakia, which ended the communist government and the establishment of a democracy.

We use the unique traits of human power I have described throughout this book in our rebellion and acts of defiance. Our love – "seeing of the other", for example, our memories, feelings, and emotions. This kind of human power is the foundation, so to speak, of human defiance. Historically, we have seen human beings' defiant acts projected precisely by these traits – powerful emotion, love, passion, resistance to injustice, selflessness, and courage. Pakistani Malala Yousafzai is an activist and advocate for girls' education. She became a global symbol of resistance against extremist ideologies when, on 9th October 2012, she and two other girls were shot by a Taliban gunman targeting her for her activism. Fearlessly, she continued her advocacy and received the Nobel Peace Prize as the youngest person in history for her struggle against the "suppression of children and young people and for the right of all children to education".[15]

Does an AI system feel injustice?

Defiance will often emerge as a resistance to external forces that oppress human power. The violent aggression towards human beings, murder, loss of life, beating, deprivation of liberty – or just the mere repression of our creativity, emotion, wisdom, and lust. Is the unruly child in school just "disobedient" to the adults who want it to conform to their norms and world views, or is it not expressing its most human individuality and uniqueness, its human power to resist?

In J. D. Salinger's novel *The Catcher in the Rye* from 1952, the young man Holden Caulfield moves aimlessly around in the city of New York – the Edmont Hotel, Central Park, Penn Station, the Natural History Museum, The Rockefeller Center skating rink, the Metropolitan Museum of Modern Art – meeting and staying with random people to avoid going to his parents' house and facing their disapproval of yet another school failure. The 16-year-old boy has been expelled from the latest prep school that he attended. Disengaged, rebellious, and failing his classes, he generally refuses to conform to the dominant social norms. Holden describes his teachers and fellow students as "phonies" only adhering and performing with values and the objective he does not see himself in, which is to get rich, as he says. There is no point in attending the school. Holden is described as a highly intelligent, reflective, curious, and creative – in language, behaviour, thoughts, and soul – young person. However, he is a child lost in the conservative values and norms of the 1950s US prep school system. The schools he attends fail to catch his attention. He needs something else.

Interestingly, just like Holden was expelled from the school, so was the book about him. In the 1960s–1970s, *Catcher in the Rye* was challenged and banned in school districts all over the US, starting with one teacher fired in 1960 for assigning the book to an 11th-grade English class.[16] The reasons were many and similar to those attached to the protagonist of the book, Holden himself – from having excess vulgar language, sexual scenes, moral issues, and excessive violence to dealing with things like "the occult" and "communism". Holden and the book about him were considered defiant in this system, a rebellion that must be expulsed. Nevertheless, today, the book is regarded as an essential critique of a repressive social system that fails to meet and reinforce the younger generation's human capacities.

Does an AI system ever resist its own education/training (data)? Does it need to?

For reasons like this, the American author John Caldwell Holt was a proponent of homeschooling. When children are not allowed to express themselves freely when requested to perform in haste, punished, and not rewarded for their curiosity and creativity, they become limited versions of the human beings they could have become, he argues:

> What are the results? Only a few children in school ever become good at learning in the way we try to make them learn. Most of them get humiliated, frightened, and discouraged. They use their minds, not to learn, but to get out of doing the things we tell them to do–to make them learn. In the short run, these strategies seem to work. (….) in the long run, these strategies are self-limiting and self-defeating, and destroy both character and intelligence. The children who use such strategies are prevented by them from growing into more than limited versions of the human beings they might have become.[17]

Thus, the most human thing to do for children to endure, he says, is to rebel. Like the boy Holden Caulfield, the children that John Caldwell Holt is writing about repress their human qualities and powers as a form of resistance to the "inhuman" education system:

> The stubborn and dogged "I don't get it" with which they meet the instructions and explanations of their teachers – may it not be a statement of resistance as well as one of panic and flight?[18]

Holt advocated for new approaches to child education, a more respectful acceptance of the value of the child's perspective and innate powers. Rebelling against what he sees as non-loving modes of education that do not want to see the human powers of the child:

> Gears, twigs, leaves, little children love the world. That is why they are so good at learning about it. For it is love, not tricks and techniques of thought, that lies at the heart of all true learning. Can we bring ourselves to let children learn and grow through that love?[19]

Holt also believed that the disobedience and resistance of the student may be seen positively as a form of human self-expression and development. In the education system, he was concerned about, children were not treated as human beings but just as "intelligent beings", he said; like intelligent machines, we improve by filling it with information. He asked:

> Is a baby nothing more than a collection of neural pathways for us to stimulate?

No, they are not, he answered, inviting the teachers and school system to respond to children with "intellect with heart".[20]

What kind of intellect does a machine use to respond to the human being's prompts?

In 1989, the United Nations Convention on the Child was adopted. The Convention's perspective on the child was, at the time, new: Children are not just unfinished, imperfect adults; they are human beings with their own rights that must be protected – rights, for example, to participate and express themselves freely. At the heart of human rights laws and conventions is the concept of "human dignity". Human dignity is a unique value of humanity that all human beings share; that which is left when we take away class, gender, race and age. "Got life, I got my life", as Nina Simone sang.

Throughout history, when our shared human dignity was challenged, humans of all ages, genders and races have rebelled and resisted with defiant acts. It is only human nature to do so. Thus, we have also established the international human rights system to protect human beings' power to defy, be critical, have their own opinions, and voice them to question the exercise against the arbitrary power that represses human power. In fact, the healthiest of democracies thrives on human critique and difference of opinion.

Why can an AI system not have human rights?

The instinct to rebel is shared among humans. The freedom to do so through individual choice is our human power. We express our unique human perspectives and histories through critical free agency. When we resist, defy, and formulate our concerns and rejection into action and communication to effect change, we express our human uniqueness. This is how Hannah Arendt understands political agency, which is indeed an individual freedom, nevertheless strongest when expressed as collective action developed in interaction with other people.

To Arendt, our individual agency is not inherent; it is deeply influenced by our interactions with other people and the character of the public sphere.[21] Individual political agency thrives in authentic public spaces that allow diversity, free-thinking, speech, and action, and it withers in politically reductive spaces when the coming together of individuals does not empower political agency through meaningful interaction and decision-making. This is, in essence, the difference between power and violence. When people are violently coerced into modes of interaction and being, when our human power to resist, revolt, and defy is repressed, our humanity is fundamentally endangered.

Who decides which conditions are unjust in an AI system?

In Part 1 of the book, I asked the following question: Could it be that digital transformations, AI, and big data are eroding the qualities of public space – the conditions for human power? I referred to the "socio-technical infrastructures of power"[22] that condition our lives, and Hannah Arendt's description of the conditions in which human political agency strives or fades.

Undoubtedly, we can consider technology not only as tools of empowerment or disempowerment but also as the very fabric of the spaces in which we live and act out acts of defiance and rebellion. These digital fabrics seamlessly merge into the backdrop of our daily existence, often going unnoticed. Nevertheless, they are crucial in structuring and streamlining our everyday life and agency. When we feel observed, we alter our behaviour. We become less spontaneous in our actions and self-expression and are reluctant to stand out or deviate from established norms. Edward Snowden's revelation of the US National Security Agency's online surveillance infrastructure, for example, had a visible impact on minority groups' expression of political views, studies show.[23]

I have elsewhere described an applied ethics approach, "data ethics of power"[24] that takes the form of "spaces" of critical negotiation that are aimed at making power dynamics of the big data and AI composition of human life and society visible and to point to alternative, more humanly empowering realities. I here argued that these "spaces of negotiation" come into existence when various "systems" – material, immaterial, technological, cultural, etc. – collide and give rise to human conflicts and disputes. The result is negotiation and, finally, collective action. This form of collective ethical agency depends on what I referred to as "Critical Cultural Moments" that I also briefly introduced in Chapter 1. These moments have distinct human characteristics and emerge when human memory and intuition are prioritised and "human power" flourishes.[25] In more practical terms, this translates into a prioritisation of the human dimension, human interests, and actively involving human beings in the design, development, adoption, and governance of big data and AI systems.[26]

Let me explain what this means with an example. In 2021, during a trip to the World Exposition held in Dubai, I visited one of the country pavilions. An eager man from the pavilion was proudly showing us around in a place with all the latest technological developments on display: A looping 53-metre-long film display of nature and technology; animated, interactive games and programmable light shows; a coffee machine based on another machine capturing CO_2 from the air combining it with hydrogen and then converted to synthetic natural gas. At one point, our enthusiastic guide brought us into an elevator to reach the second floor. It looked like any other elevator with the usual up and down buttons. However, when I went for the button to bring us up, the guy patiently explained the wonders of the elevator to me: "You can use voice recognition. Tell it to go up". And so, we stood there, a small group of humans, in the elevator, prompting in different voices and sound levels, an elevator to bring us up. It didn't budge for a while, but finally, it recognised one of the humans' voices and brought us up the ten steps of a standard staircase to the second floor. Technology can certainly make you feel empowered but also disempowered on all kinds of levels. Oddly enough, the little group of people, including myself, did not reach out for the button that would have brought us to the second floor immediately; we stood patiently still in the tiny, cramped space of the elevator with voice recognition asking it in all kinds of polite ways to bring us to where we needed to be. Why didn't I push that button? Or walk the ten steps up the stairs?

There are countless examples of disempowering socio-technical constellations that condition our human critical agency. Many with much more severe impacts on human critical agency than an elevator that refuses to comply. Some are more obvious than others. Clearly, a social scoring credit system like the ones used in some Chinese cities has chilling effects on a citizen's willingness to voice controversial opinions or act in any way that may be considered "disobedient" by the Chinese government. For example, in the small city of Rongcheng, China, every resident is given a base personal credit score of 1000. This will then be influenced by the citizen's good or bad acts. These could include voicing critical opinions online. As Zeyi Yang recalls in MIT Tech Review, although later overruled in a 2016 rule, the city once decided that "spreading harmful information on WeChat, forums, and blogs" would subtract 50 points from your score. She gives the example of a case where one resident lost 950 points for distributing letters online about a medical dispute.[27]

Many disempowering technologies are less obvious. If you are in my generation or previous generations, you might remember a life without social media and clearly the difference between then and now – that you had different arenas of interactions, other ways of retrieving information, different ways of engaging with the world back then. The importance of the internet in your life today is obvious. No one would dispute that access to knowledge, for example, is much easier. But you might not be aware of the analytical scrutiny you are at the same time subjected to when engaging online with the platforms that shape what and how you access information and engage with the world and other people.

Scholar Nathalie Smuha discusses the impact of social media and their AI analytical capacities on the collective public sphere and discourse that Hannah Arendt championed as the foundation of political agency.[28] She says that social media initially carried great promises for interpersonal engagement and collective action. Yet, increasingly "surveillance by default" – business models have hindered this potential. The AI and data-driven systems prioritise extreme and polarising content and reinforce existing world views and opinions when creating echo chambers of information around users. In this way, public discourse is fragmented, and powerful actors, such as politicians, may invisibly micro-target citizens with tailored messages and undermine accountability. The result, Smuha argues, is a polarised and fragmented public realm challenging the sense of solidarity and unity among people that Arendt considered necessary for collective action.

Initially, as I have also described it in Part 1 of the book, the Internet and the World Wide Web were developed with very different ideas about their potential for human dissent and political agency. For example, in their respective "cyber" manifestos, John Perry Barlow and Donna Haraway envisioned cyberspace as free from the restraints of traditional forms of power. And there are, of course, many examples in recent history where social media platforms were key tools for citizen dissent, useful when people have gathered to voice and act collectively on shared political concerns.

In the early 2010s, thousands of citizens gathered across the Arab world for anti-government protests. Protesters used social media to mobilise, organise, and share their protests and experiences with the rest of the world to gain support.[29] Indeed, technology can also empower human defiance and collective rebellion. But it depends on how it is designed.

Does a machine defy and rebel? Will it ever?

Social media that by design lack privacy and track users, can be used by undemocratic governments to identify and control dissidents. We can certainly think of this as a disempowering design that counter human defiance. However, we should also try to imagine an alternative design of these platforms that empowers those that are subject to brutal power to resist unjust conditions without risking their safety and lives. It is, evidently, not a technology design that will emerge in the current dominant online industry on its own. Other more inclusive design spaces for an alternative digital design are fortunately evolving. An example is the "Design Justice Principles" created by a group of independent designers, artists, technologists, and community organisers to inspire collaborative, inclusive design processes asking an important question for humanly empowering digital design:

> How could we redesign design so that those who are normally marginalized by it, those who are characterized as passive beneficiaries of design thinking, become co-creators of solutions, of futures?[30]

NOTES

1. Thank you, Clara, for your thoughts on the movie.
2. "Albert Camus Facts", https://www.nobelprize.org/prizes/literature/1957/camus/facts/

3. Camus, A. (2000) *The Rebel* (translated by A. Bower), Penguin Books Ltd. (p. 10) (originally published in French in 1951).

4. Camus (2000/1951, p. 11).

5. Camus (2000/1951, p. 10).

6. Marx, K., Engels, F. (1969) *Manifesto of the Communist Party* (translated by S. Moore), Progress Publishers (original work published in German in 1848), https://www.marxists.org/archive/marx/works/1848/communist-manifesto/

7. Adorno T. W., Horkheimer, M. (1977) "The culture industry: enlightenment as mass deception" in J. Curran et al. (eds.), *Mass Communication and Society* (original published 1944), Edward Arnold.

8. Foucault, M. (1991) *Discipline and Punish: The Birth of a Prison.* Penguin (originally published in French in 1975).

9. Fanon, F. (1963) *The Wretched of the Earth* (p. 100) (translated by C. Farrington), Grove Press (original work published in 1961).

10. Fanon (1963, p. 37).

11. Hao, K., Swart, H. (2022, April 19th) "South Africa's private surveillance machine is fueling a digital apartheid", *MIT Technology Review.* https://www.technologyreview.com/2022/04/19/1049996/south-africa-ai-surveillance-digital-apartheid/

12. Hao, K., Seetharaman, D. (2023, July 24th) "Cleaning up ChatGPT takes heavy toll on human workers", https://www.wsj.com/articles/chatgpt-openai-content-abusive-sexually-explicit-harassment-kenya-workers-on-human-workers-cf191483

13. Guo, E., Renaldi, A. (2022, April 6th) "Deception, exploited workers, and cash handouts: How Worldcoin recruited its first half a million test users", https://www.technologyreview.com/2022/04/06/1048981/worldcoin-cryptocurrency-biometrics-web3/

14. Thoreau, H. D. (1849) *On the Duty of Civil Disobedience*, https://www.gutenberg.org/files/71/71-h/71-h.htm

15. "Nobel Prize Laureates by Age", https://www.nobelprize.org/prizes/lists/nobel-laureates-by-age/

16. "Banned and/or challenged books from the Radcliffe Publishing Course top 100 novels of the 20th century – *The Catcher in the Rye*, JD Salinger", American Library Association, https://www.ala.org/advocacy/bbooks/frequentlychallengedbooks/classics

17. Holt, J. C. (1995) *How Children Learn* (originally published in 1967), Da Capo Press.

18. Holt, J. C. (1995) *How Children Fail* (originally published 1964), Da Capo Press.

19. Holt (1995/1967).

20. Holt (1995/1967).

21. Thuma, A. (2011) "Hannah Arendt, Agency, and the Public Space", in M. Behrensen, L. Lee & A. S. Tekelioglu (eds.), *Modernities Revisited, IWM Junior Visiting Fellows' Conferences Proceedings*, Vol. 29.

22. Hasselbalch, G. (2021) *Data Ethics of Power – A Human Approach in the Big Data and AI Era*, Edward Elgar.

23. Stoycheff, E. (2016) "Under surveillance: examining Facebook's spiral of silence effects in the wake of NSA internet monitoring", *Journalism & Mass Communication Quarterly*.
24. Hasselbalch (2021).
25. Hasselbalch (2021, pp. 126–127).
26. Hasselbalch (2021, p. 5).
27. Yang, Z. (2022, November 22nd) "China just announced a new social credit law. Here's what it means", *MIT Technology Review*, https://www.technologyreview.com/2022/11/22/1063605/china-announced-a-new-social-credit-law-what-does-it-mean/
28. Smuha, N. A. (2022) *The Human Condition in An Algorithmized World: A Critique through the Lens of 20th-Century Jewish Thinkers and the Concepts of Rationality, Alterity and History*, Institute of Philosophy, KU Leuven.
29. Smidi, A., Shahin, S. (2017) "Social Media and Social Mobilisation in the Middle East: A Survey of Research on the Arab Spring", *India Quarterly*, 73(2), 196–209.
30. Design Justice Network (Living document 2018), *Design Justice Network Principles*, https://designjustice.org/read-the-principles

CHAPTER 7

Wisdom

Is a Machine Wise?

A man is locked inside a room. Underneath the door, a piece of paper slips with something written on it. The problem is that it is in Chinese, and the guy doesn't know Chinese. How does he respond to this? He looks around and finds a computer with a programme that he can follow to place Chinese characters in an order that responds in a structurally correct manner, answering the string of characters on the sheet of paper. He can now respond to the sender of the sheet of paper outside the room and convince this person that he can speak Chinese. But does this mean that he really understands the language? Can the person outside the room trust that he communicates with a real-life human being?

In his famous "Chinese Room" paper published in 1980, philosopher John R. Searle argues that he should not think this is so.[1] Searle himself, who is the man he in this example imagines is locked inside the room, still does not have the slightest idea of what the Chinese characters on the sheet of paper mean or even what he responded to them. Searle, therefore, concludes that "no program by itself is sufficient for thinking".[2]

In 1950, the mathematician and computer scientist Alan Turing famously proposed a method for testing a machine's ability to display intelligent behaviour indistinguishable from a human. He called this the

DOI: 10.1201/9781003527855-9

"imitation game" that was to prove a computer's "thinking" capacities. In the game, one human would interrogate a computer designed to exhibit human-like behaviour and a human being in another room via a text-based system (a "teleprinter"). If the interrogator was convinced by the text exchange with the computer that he was communicating with another human being, the computer could be said to be "thinking".[3] Searle's Chinese room example, on the other hand, shows that Turing's test is based on the false assumption that human intelligence is like the computer's information processing programme. Searle argues that this test does not prove that the computer programme "thinks" like a human because it does not understand like a human. It is not "intelligent" like a human.

Does a machine think?
Does it understand?

What is human intelligence? What is thinking? How do we evaluate and reward intelligence? These questions have always been central to the short history of computing for artificial intelligence (AI). Nevertheless, answers were never honestly sought answered. The idea that we can programme a machine to think like the human mind is based on a certain stubborn confidence in what the human mind is, what it excels at and what it certainly cannot do.[4] Thus, since the 1950s, scholars have been more focused on investigating the similarities between the human brain and computers during various periods of advancement in AI science rather than achieving a greater understanding of what the human mind and thinking indeed entail.[5] Hence, human thinking capabilities were reduced to what is understandable to a computer: data – and data processing.

In Turing's test, the human mind is conceived of as a programme that processes information convincingly. Others like Norbert Wiener, the father of Cybernetics, have claimed similarly that human neural biology is a form of communication like any information processing system. It is complex but no different than the data processing of a machine.[6] Along these lines, philosopher Luciano Floridi describes the world and everything in it as an information environment, an "infosphere" reinforced by digital and ICT technologies, and humans as information processing organisms, "inforgs" on par with other artificial agents.[7]

We also see this kind of scientific manoeuvring with existential concepts in the development of computer programmes like chess-playing systems equipped to formulate and execute strategic moves in games to compete with human chess players. And by now, computers have also, on several occasions, beaten humans in games like these, like when IBM's Deep Blue computer won over Russian Chess Grandmaster Garry Kasparov. Some have seen these events as proof that we are finally on our way to the ultimate artificial general intelligence (AGI) moment when computers outcompete human intelligence once and for all. We also see these ideas expressed in the mission statements of leading AI companies today on how AI is outperforming humans, reaching and even outcompeting the human potential. As when OpenAI's CEO Sam Altman, in the blog post "Planning for AGI and Beyond", published in February 2024, wrote about their mission:

> (…) to ensure that artificial general intelligence – AI systems that are generally smarter than humans – benefits all of humanity.[8]

Altman did not bother to define in the post what he meant by "smarter than humans" other than it has something to do with helping humans with "any cognitive task", make "the rate of progress in the world much faster", and "accelerate science". His assumptions about the human mind and human intelligence, like many other AI company leaders, researchers, enthusiasts, and even critics today occupied with the idea of AGI, is that a machine that can outperform a human in information storage, analysis and processing at an incredible speed, is essentially smarter than a human. Of course, if this is how we perceive human intelligence – from the point of view of what a computer excels at – not much is left to interpretation regarding the potential and uniqueness of the human mind, thinking, and intelligence.

The presentation of the potential AI power compared to human power is in the AI-hyped debate today, however, incredibly simplistic compared to the reflections made on the topic in early computer history.

The mathematician Ada Lovelace, known for her writings on the early general-purpose computer ("The Analytical Engine"), for instance, famously said:

> The Analytical Engine has no pretensions whatever to originate anything. It can do whatever we know how to order it to perform.[9]

Even, Alan Turing, who had a great interest in the development of computers that could imitate the human intellect and even believed that they would at some point supersede our basic thinking capabilities in speed and storage, still emphasised that the development should stop there and not try to imitate other human traits:

> I certainly hope and believe that no great efforts will be put into the making of machines with the most distinctly human, but non-intellectual characteristics such as the shape of the human body; it appears to me to be quite futile to make such attempts and their results would have something like the unpleasant quality of artificial flowers.[10]

Thus, the "thinking" capabilities of a computer, AI, he warned, should not be confused with other traits of human power:

> Attempts to produce a thinking machine seem to me to be in a different category.[11]

Let me start there. Much of the AI industry's narrow and vague ideas about human capacities today are limited and based on a very selective approach to original ideas about computers, AI, and humans. As a point of departure, let's therefore turn to other sources in our rich human fields of knowledge and imagination about the constitution of "human intelligence".

First, consider human intelligence as something more than Turing's computer, that is, more than just useful information that we store and process in efficient ways. Let's consider human intelligence a trait of human power that we may call "wisdom". Going back through history, the concept of human "wisdom" has been subject to wide scientific, artistic, and philosophical speculation.

Aristotle and Plato expressed various views on wisdom as an ideal, a specific "godly" kind of intellectual love that earthly human beings may cultivate and strive for.[12] But even something as simple as a basic entry on wisdom in a dictionary like the Merriam-Webster Dictionary, although undoubtedly also biased (I'll get back to that), still expresses a more creative and nuanced understanding of what constitutes a wise being:

1. a: ability to discern inner qualities and relationships: insight
   b: good sense: Judgement c: generally accepted belief (…)
   d: accumulated philosophical or scientific learning: knowledge
2. a wise attitude, belief, or course of action
3. the teachings of the ancient wise men[13]

Is a computer wise when it processes vast amounts of information, makes a complex analysis and proposes a useful decision?

We can start by exploring depictions of wise human beings in history. What do all these people have in common? Why do we see them as wise "thinkers" or "intellectuals"? Indeed, we do not consider them wise just because of their knowledge and the information they possess and manage to process in "intelligent" and useful ways. There is more to wisdom than that.

The Italian painter Sandro Botticelli's *The Birth of Venus*, painted in 1485 amid the Italian humanist Renaissance, represents a vision of human wisdom. Venus – the Goddess of beauty and love – stands shyly covering her earthly naked body with her long hair on an unnaturally large seashell, on one side blown towards human shores by two goods and on the other side received by another god with a blanket to cover her naked body. Venus is a complex image of human wisdom, of two types of human attraction – the physical ("earthly") and the intellectual love ("heavenly"); that which we have, our physical and very material minds and bodies, and the kind of intellectual beauty, or "wisdom", that we strive for.

Another image we have of human wisdom comes from the descriptions of the wise woman Hypatia. She lived and was brutally murdered 1600 years ago in the intellectual centre of the ancient world, Alexandria. In history, she stands out as an exceptional intellectual, educated by her father, the philosopher Theon of Alexandria, who was also the director of the Museum of Alexandria.[14] None of her work is available anymore; no scientific and philosophical information has been preserved. But we know about her from various descriptions of her tragic story when she was murdered and torn to pieces by a mob of people, and we know about her character and her world-famous teaching style.

In the tenth-century Byzantine encyclopaedic dictionary, her "exceptional wisdom" is, for example, described as a complex of an attractive kind of inner and outer beauty that both entices envy and love and admiration of others. She is highly knowledgeable in mathematics, philosophy, and astronomy. Still, she is also said to have had a noble character and possess a particular kind of practical virtue:

Skilful and eloquent in words and prudent and civil in deeds.[15]

This kind of wisdom is expressed in her excellent teaching skills, which are outstanding to the extent that students would travel from all corners of the world to attend her classes.[16]

In this depiction of the wise woman Hypatia we already see wisdom as something other than the knowledge we produce. A wise person like Hypatia is known as such, first and foremost, because of her character and, in particular, her ability to teach others. This is what stands out as the most profound character of her wisdom:

> All the contemporary and later writers of this period testify to the high reputation of her work as a teacher. Each one attributes an extraordinary eloquence and an agreeable discourse to her lectures. Suidas speaks highly of her teaching methods, while Synesius in one letter praises her voice and in another mentions that her philosophy was carried to other lands.[17]

How does an AI system teach? How does it cultivate knowledge?

The Greek philosopher Socrates' views on wisdom as a process of education and cultivation are well known. He famously held that the very awareness of the incompleteness of knowledge makes a human being wise.[18] All humans can do is strive for wisdom. We also recognise this idea about human wisdom as the core principle, "Humanitas", of the 15th–16th-century humanist idea of the cultivated individuals' inherent attributes and virtues that make them capable of governing their existence with empathy and compassion towards others.

Does a computer programme know that its information is not complete?

We can think of traditional forms of education as cultivation towards human ideals of wisdom. In school and university, students are taught according to curricula that specify what is considered most important and relevant of the accumulated human knowledge on a subject that the student should ideally learn. The Buddhist leader of the Tibetan people, the 14th Dalai Lama, Tenzin Gyatso, also describes other forms of cultivation for wisdom, such as the "inner development" of the human being and the cultivation of our positive and negative emotions through, for example, meditation. He holds that negative, destructive emotions restrain our inner freedom and thus "impair our judgment". Wisdom, to him, is freedom from negative emotions that get in the way of "natural and spontaneous compassion",[19] or as Renuka Singh explained it to me:

> Wisdom is a state of mind where all negative emotions have melted away like snow on a mountain. It means emptying yourself entirely so that you can receive. If the cup is full, you can't pour another drop of tea, coffee or milk into it. But if the cup is empty, you can keep filling it. Knowledge is like that.[20]

Where does a machine's "knowledge" come from?
What kind of knowledge does an AI system cultivate?
Can an AI system receive the knowledge of the universe?

To be wise is to be free. Human wisdom is a freedom, a confidence to make judgements and choose freely without external pressure, or a "full cup". As we've seen in other chapters of this book, "free thought", a human inner intuition or confidence, is at the core of human power. Hanna Arendt wrote about human contemplation and an uncorrupted public space as a condition for true political thought. Jean-Paul Sartre talked about free choice as the foundation of moral thinking and, thus, as a type of human responsibility.

In depictions of wise human beings in history, we see that they exhibit not only vast amounts of knowledge but also confidence and the agency and freedom from external pressure to make decisions based on a complex of emotions, intuitions, memory, situation, and knowledge. Dalai Lama himself we conceive as a person of great "wisdom", but this perception does not go without reference to the confidence he exudes. As in the story of the meeting between Dalai Lama and the French Nepalese writer, photographer, and Buddhist monk Matthieu Ricard, who "...felt overwhelmed by the master's peaceful compassion, wisdom, and mountain-like inner strength".[21]

We see this kind of unbending confidence in the lives and actions of wise people like the Indian spiritual teacher Jiddu Krishnamurti who spent his life travelling around the world speaking to groups of people and individuals about life, human existence, and politics. He was a sought-after speaker, gathering crowds wherever he went and forming intellectual alliances with influential thinkers such as Indira Gandhi, Fritjof Capra, and Aldous Huxley.[22] After many years as part of the theosophical organisation, he received an advanced spiritual position, but then, surprisingly, he rejected it and withdrew from the organisation. Considered one of the time's most essential and admired gurus, he, in contrast, said:

> There's no guru, no teacher, no saviour, nobody outside that can bring about this extraordinary state of harmony.[23]

Is an automated decision-making system free? Is it confident?

Jiddu Krishnamurti also said that "Wisdom is the ending of suffering...".[24] We perceive human wisdom as liberty and confidence, but not without a strong ethical dimension, a social dimension expressed in humility, compassion, and a sense of responsibility towards others. We cannot deny that Adolf Hitler was intelligent. Many other brutal dictators have managed their regimes and navigated world powers with cunning intelligence. However, we do not consider these human beings "wise". The humanist principle "Humanitas" does not imply the cultivation of just any kind of knowledge but, above all, human moral development.

The Kenyan scholar and political activist Wangari Muta Maathai received the Nobel Peace Prize in 2004. After studying biology in the United States, she returned to Kenya and, in 1971, became the first East African woman to receive a PhD. Wangari Maathai started the Green Belt Movement in 1977 to halt deforestation in Kenya by encouraging women to plant trees and establish tree "nurseries".[25] These women were disadvantaged in a patriarchal society and culture. Wangari Maathai encouraged them to take responsibility for their local natural environments and empowered them in their communities. Today, more than 50 million trees have been planted, and the movement has inspired ecological thinking worldwide. "Be a hummingbird", Wangari Maathai said, alluding to the power of the wisest individual who makes differences in their communities by themselves when they are most often not the most powerful in a bigger crowd. The hummingbird in the story she used to tell about the power of the few is a tiny bird in a burning forest that, with the greatest force, flies back and forth with tiny drops of water to slow down the fire, ignoring the mocking by the other much bigger and resourceful animals.[26]

Wangari Mathai also shows us that wisdom cannot exist without human memory, culture, and experience:

> Wisdom is coded culture (…) that has been accumulated for thousands of years and generations. Some of that wisdom is coded in our ceremonies, it is coded in our values, it is coded in our songs, in our dances, in our plays.[27]

Wisdom is the collective knowledge of our human history and culture, and it is the memories of a human lived experience. Depictions of wisdom will most often refer to human history, age, and experience.

In Carl Emil Doepler's painting "Odin, der Göttervater" (1882), the Viking God Odin sits on his throne, with his old hands firmly on the armrest, an aged face with a long white beard, and his two ravens, Hugin and Munin. In the engraving "The Wise King Solomon" (1866) Gustav Doré has depicted King Solomon, who asked God when he appeared to him in a dream for "understanding to discern justice" and was then given "a wise and understanding heart".[28] The engraving shows once again a very old man with a long white beard, clear signs on his face and hands of the years he has lived and the experiences he had, the pen and scroll in his hands on which his wisdom materialises in letters. Depictions of wisdom as this often illustrate this complex of a biologically lived human experience and memory and its external cultural expression in books and scrolls.

In fact, we can think of human wisdom as a complex of two forms of memory and experience – one that is represented and can be preserved and processed, another that is lived, cannot be represented or even revisited. In this way, Henri Bergson also famously described the time of our clocks as a mechanical framework we have invented to quantify, divide, and structure our time and make society work,[29] and the very different continuous, uninterrupted time, "duration", that we can only truly grasp while experiencing it:

> Time is invention or it is nothing at all.[30]

Our books, films, computers, and libraries represent the first kind of time. They can store and preserve our human collective memories, and some of these tools of cultural preservation can even process and make sense of it. The library is, for instance, a monument of human wisdom; we can mourn and suffer the loss of it like we regret the destruction of the library of ancient Alexandria. It is almost unbearable to think of the loss of the thousands of scriptures and books of human wisdom from all over the world that perished due to war and other forms of human neglect and destruction. None of Hypatia's texts have been salvaged over time; most likely, they were destroyed together with the Library of Alexandria.

Nevertheless, human wisdom endures due to the persistence of human memory that transcends its representation. Salvador Dali's painting *The Persistence of Memory*, which shows melting clocks intertwined with a melting human face, illustrates a wisdom like this – detached from and even subversive of the mechanical memory of "clock time". It persistently melts the clock because it is more than just a clock. It is the persistence of human time.

With Bergson we may perceive a great potential in the human lived experience and memory, "duration", through which we access and perceive continuous, uninterrupted time. This is not to say that we do not need to represent and preserve human wisdom in a collective form that makes sense to more than just the living individual. Computers are great for a preservation task like this. They certainly have more significant data memory and storage capacities than humans. According to ChatGPT, OpenAI's GPT-3 had, in 2022, 175 billion parameters, making it one of the largest language models on the market with immense requirements for storage, both during the training and the deployment of the model. Thus, this is a machine that can process an abundance of culturally coded content. Much more than a single human being ever could. Does this mean, then, that an LLM like ChatGPT is as wise as the collective human culture it holds?

Is the memory capacity and storage of a computer like the memory of a human being?

Let me answer this question with yet another example. When his wife is murdered and he is knocked out by an assailant, Leonard Shelby, the main character in Christopher Nolan's movie *Memento* (2000), is left with Anterograde Amnesia. This is a condition that prevents him from making new long-term memories. His memories from before the incident are still intact, but Leonard cannot create new memories and forgets everything he experiences shortly after it happens. He, therefore, takes Polaroid photos, scribbles down notes, and tattoos his body with clues for his present self, who is obsessed with only one thing – to solve the murder of his wife. Leonard experiences everything in instants; his memory of the past is no longer continuous but externalised in data on the notes, pictures, and on his body that he uses only for the present purpose of solving the murder mystery. He is constantly "waking" up confused, trying to piece together his memories. Left with a fragmented experience of time and his own self, where the past – his memories – are no longer an experience but a set of data utilised only for what to him often appears as a non-sensical present purpose. Because, of course, solving the murder of his wife does not make any sense when his memory of this is separated from his immediate experience. When he finally, at the end of the movie, solves the mystery, he moments after is once again as lost as ever and in more than one sense, which his final words of the movie eloquently illustrate: "Now where was I?"

AI systems, like the Large Language Models (LLMs) we increasingly adopt into our everyday lives, cannot function without data and memory. Their intelligence is derived from storing and processing vast amounts of data. Just like it is for humans and human culture. That said, while AI systems certainly have better capacities for storing the data and reminders they need, they are also very similar to Leonard Shelby's approach to this "historical data". Like Leonard, they do not have continuous memories; they only have fragmented representations that they piece together and repurpose for present tasks. Think of the experience you have when prompting an LLM to produce knowledge. There is no genuine "I" in the response, no continuity of time, and no experience. The temporal existence of an LLM is instantaneous and vacuumed into the present. On the other hand, the memories from which human wisdom springs are complexes of personal, cultural, historical, and, importantly, lived experiences. Indeed, books and computers can undoubtedly support and improve human wisdom, but they cannot be said to be wise in and by themselves.

Thus, human wisdom is also a unique perception of and approach to time. A wise human being doesn't just gather and store information in memory for later use. Human wisdom involves a continuous movement that connects time and space. Wisdom entails an awareness of the interconnectedness of life, cultures, people, and societies. This is also how Dalai Lama describes wisdom:

> You always have to approach from a multitude of perspectives, a variety of mental factors, a variety of understandings. It is more complex than simply There's the problem and here's the antidote.[31]

What is the difference between human wisdom and machine intelligence?

Returning to the wise Viking god Odin, he is perhaps also sending us a wise message through time. He has only one functioning eye; the other he has sacrificed to gain an unconceivable amount of knowledge. But he has his two ravens, Hugin (thought) and Munin (memory). They see and hear everything; they can talk, remember all, and predict the future. Odin depends on them, letting them roam wild to scout the world for him, but this is also a trade-off, a delegation of his powers that he has to accept to be able to control the present and see the future. And so he also frets:

> Hugin and Munin fly each day over the spacious earth. I fear for Hugin, that he come not back, yet more anxious am I for Munin.[32]

These concerns of an ancient Viking god spell out a deep human anxiety that we urgently need to revisit today. Anxiety about losing something fundamental to human wisdom when automatic systems for storing and producing knowledge are introduced into our everyday lives. Which trade-offs are we willing to accept in our yearning to surpass the limits of the human body and mind when developing and adopting AI? Here, we might learn from Odin's anxiety about the potential loss of memory (Munin) because what is thought (Hugin), an intellect – what is a machine that thinks – without the situated dynamic qualities of human memory (Munin) and experience?[33] Leonard Shelby might tell us the answer to this.

Is a machine wise?

To summarise, machines can be intelligent but never wise – quite the contrary. The wisest people throughout history are recognised for the traits that most profoundly distinguish human beings from machines: Love, situated memory, intuition, feeling and emotion, life, and defiance. These traits are, in essence, the wisdom of humankind and the human expertise that our technology politics should aim to defend, as Professor Frank Pasquale urges us to do in his book *New Laws of Robotics*.[34]

Indeed, human power is a wisdom that has empowered humans through history to create unique and extraordinary things. It must be safeguarded, reinforced and defended, not diminished, mocked, and disrespected. As former director at the Council of Europe, Jan Kleijssen summarises it:

> Human power is human genius. It's the Leonardo da Vinci's, it's the Michelangelo's. It's the unique spark that brings human power. Whether you're religious or not, it's the 16th chapel in Rome, where you see God touching the finger of Adam and having a divine spark. And for me, that is human power. It's what's given us the Colosseum, which has given us all the beautiful paintings and artworks of the world. The beautiful music, the Mozarts. That's for me, human power.

We could also describe human wisdom and wise decision-making as the accumulation of several of the traits of human power that I have described in this book. This is at least the way Renuka Singh describes wisdom:

> Being wise means considering the welfare of the entire universe, including every creature on Earth. It involves being in touch with your creative and compassionate self, and acting in a way that benefits everybody. This is wise decision-making.

Now, while descriptions and depictions of human wisdom in history present us with a more nuanced understanding of what constitutes the human mind and intelligence than the AI movement of the late 20th and early 21st century seem to do, they are at the same time biased as to whose wisdom is rewarded and most valued in society. After all, they are made by humans.

The Merriam-Webster Dictionary defines wisdom, among others, as "the teachings of the ancient wise men", August Rodin's famous statue *The Thinker* towers above the spectator in the form of a hugely muscled man, countless paintings show the bible's three wise men that visit Jesus. Just like the machine reproduces our biases and fundamentally limits our perception of ourselves and our potential, so do we humans, time and again, undermine ourselves by continuously reproducing human biases. For example, when exploring this chapter's topic, it wasn't easy to find depictions of wise women in human history. This makes me think that the wisest thing to do now is not to invest more energy in the futile project of creating "wise" machines but to focus on solving our real human problems, our most primal brutal urges to dominate with force and suppression.

Will AI intelligence ever outperform human wisdom?

NOTES

1. Searle, J.R. (1980) "Minds, brains, and programs", *Behavioral and Brain Sciences*, 3(3), 417–457.
2. Searle (1980, p. 417).
3. Turing, A. (2004) "Computing machinery and intelligence", in J. B. Copeland (ed.), *The Essential Turing: The Ideas that Gave Birth to the Computer Age* (p. 441) (originally published in 1950), Clarendon Press.
4. Woolgar, S. (1987), "Reconstructing man and machine: a note on sociological critiques of cognitivism", in W. E. Bijker, T. P. Hughes, & T. Pinch (eds.), *The Social Construction of Technological Systems* (pp. 311–328), MIT Press.
5. Crevier, D. (1993) *AI: The Tumultuous History of the Search for Artificial Intelligence*, Basic Books.
6. Wiener, N. (2013) *Cybernetics or, Control and Communication in the Animal and the Machine*, 2nd ed. (originally published in 1948), Martino Publishing; Bynum, T. (2010) "The historical roots of information and computer ethics", in F. Floridi (ed.), *Information and Computer Ethics*, Cambridge University Press.
7. Floridi, L. (1999) *Philosophy and Computing: An Introduction*, Routledge.
8. Altman, S. (2023, February 24th) "Planning for AGI and beyond", OpenAI, https://openai.com/index/planning-for-agi-and-beyond/
9. Quoted in Turing (2004/1950, p. 482).
10. Turing (2004/1950, p. 486).
11. Turing (2004/1950, p. 486).
12. Laude, P. (2019), "Reflections on re-learning to be human in a global age", in P. Laude & P. Jonkers (eds.), *Philosophy as Love of Wisdom: Its Relevance to the Contemporary Crisis of Meaning* (Series I, *Culture and values*; Vol. 48), Council for Research in Values and Philosophy, p. 17; and e.g. Plato (*360 B.C.E. Symposium* (translated by B. Jowett), https://classics.mit.edu/Plato/symposium.html: "For God mingles not with man; but through Love. all the intercourse, and converse of god with man, whether awake or asleep, is carried on. The wisdom which understands this is spiritual; all other wisdom, such as that of arts and handicrafts, is mean and vulgar."
13. Merriam-Webster, "Wisdom", in *Merriam-Webster.com Dictionary*. https://www.merriam-webster.com/dictionary/wisdom
14. Richeson, A. W. (1940, Nov.) "Hypatia of Alexandria", *National Mathematics Magazine*, 15(2), 74–82.
15. Suda on Line. *Upokrisis (upsilon 166)*. http://www.stoa.org/sol-entries/upsilon/166

16. Richeson (1940).
17. Richeson (1940, pp. 79–80).
18. Plato, 360 B.C.E.
19. Goleman, D. (2003) *Destructive Emotions – How Can We Overcome Them? A Scientific Dialogue with the Dalai Lama* (p. 86), Bantam Dell.
20. Interview 2024
21. Goleman (2003, p. 74).
22. https://jkrishnamurti.org/
23. Krishnamurti, J. (n.d.). *Religion and Meditation.* https://jkrishnamurti.org/content/religion-and-meditation
24. Jiddu Krishnamurti in conversation with Allan W Anderson, San Diego, 1974, https://www.youtube.com/watch?v=9sMjDEbJz6E&list=PL8ylzMftxdLcGXuSNbWxu7tSGb7e9W8Mo
25. *Wangari Maathai Facts, The Nobel Prize,* https://www.nobelprize.org/prizes/peace/2004/maathai/facts/
26. *Wangari Maathai & The Green Belt Movement,* https://www.youtube.com/watch?v=BQU7JOxkGvo
27. Merton, L., Dater, A. (Dir.) (2008) *Taking Root: The Vision of Wangari Maathai* [Film]. Merton & Dater Productions.
28. Bible Gateway. (n.d.). 1 Kings 3 (NKJV), https://www.biblegateway.com/passage/?search=1%20Kings%203&version=nkjv
29. Bergson, H. (2004) *Time and Free Will: An Essay on the Immediate Data of Consciousness,* Taylor and Francis Group, ProQuest Ebook Central (originally published in French in 1889).
30. Bergson (1914/1907, p. 361).
31. Goleman (2003, p. 164).
32. "Grímnismál" in Thorpe, B. (1907) *The Elder Edda of Saemund Sigfusson, and the Younger Edda of Snorre Sturleson,* Norroena Society.
33. Thorpe (1907).
34. Pasquale, F. (2020) *New Laws of Robotics: Defending Human Expertise in the Age of AI,* Harvard University Press.

II

21st-Century Technology Politics

Making it Work for Humanity

One late afternoon on an unusually hot day in May 2024, I sit across the table from another human being who tells me an extraordinary story about her life and work as a composer of symphonies. Marianna Filippi, whom I've introduced a few times throughout the book, tells me she was raised in a culturally rich environment surrounded by artists, writers, and music lovers. Her early musical inspirations included a variety of musical genres from very different periods of human history – folk, Irish, Celtic, jazz, and film scores. She began playing the piano at age two and later learned the violin through Suzuki lessons, where you learn by ear, not by notes. At seven, she started creating her own folk tunes on a keyboard, and her interest in music composition developed from then on. Three of her earliest solo piano pieces are vividly present in her memory. She tells me they had a complex, thematic, film score-like quality. She would sit at her keyboard, expanding simple ideas into elaborate themes. Each piece was based on imaginary characters with dynamic emotional lives, often accompanied by stories and drawings she created to portray these figments of her imagination in music. By age 15, she started learning proper musical notation, solidifying her aspiration to become a composer.

Marianna tells me her story with sparkling eyes. Making little happy humming sounds to emphasise her points. That's just something she

DOI: 10.1201/9781003527855-10

does, which is very special about her. She moved to Denmark a few years ago from the United States to study at the Royal Danish Academy of Music in Copenhagen and was soon after commissioned by the Royal Theatre of Copenhagen and other theatres to compose music pieces. Marianna talks about her intuition (see also Chapter 4 on intuition), imagination, the flow of her creative process (see also Chapter 1 on creativity), emotions (see also Chapter 2 on emotion), and her individual experience when composing pieces of music. At one point, I ask her about her last symphony. How does she create harmonies using various instruments, musicians, talents, and sounds? She immediately has an answer:

> Some people just think about the instrument. They don't think about the human being. They think about just what this instrument can do. Like, how can I break people's eardrums?

Marianna laughs but then continues more seriously:

> I mean, they only think about how they can extend the repertoire of this instrument with all these different extended techniques. But I don't want to think about just the instrument. I want to think about the person being channelled through the instrument, about the musicians themselves, and that means stepping away from the instrument and its sound, the notes.

She pauses and is quiet, pondering for a moment, and then she picks up:

> At the end of the day, music is just black notes on a piece of paper. The coolest thing about composing is giving someone else this sheet of paper with all this music. The musicians then use themselves and their musicality to read it. The music is channelled through them, and then they play it to the audience. It's like this unspoken circuit of spiritual communication; I can't explain it, but that is actually the most amazing thing about it: that, yes, I wrote this piece, and it's coming from my own emotions, but the only way that other people can understand it is if they interpret it through themselves.

Later that evening, I found Marianna's symphony Jordens Sjæl (The Soul of the Earth) on YouTube. The Royal Danish Academy of Music

Symphony Orchestra performed it two years before at the Copenhagen Philharmonic Hall. It is an astonishing piece of music, combining and giving room for double bass, trumpet, percussion solos, and piano and harp passages. The orchestral musicians are also singing the words for "Earth" in seventeen different languages from around the world: Tellus Mater (Latin), Dedamit's (Georgian), Bumi (Malay), Erde (German), Dharti (Hindu), Dìqiú (Mandarin), Delkhii (Mongolian), Lurra (Basque), Zemli (Ukrainian), Ardhi (Swahili), Nilam (Tamizh), Domhan (Gaelic), Adamah (Hebrew), Dünya (Turkish), Gaia (Greek), Terra (Italian), and Jorden (Danish/Swedish/Norwegian).[1]

That evening, after speaking with this music composer and watching the musicians perform her symphony, everything I've spent the last three years writing this book about comes together. Her musings about her human creativity and imagination, as well as her hard training and work as a composer to make instruments, and above all, human beings, as she said, unite in musical harmonies. I see them as an allegory of the development of the technology politics I wanted to describe with this book – the one for humanity. I wonder, are we today not precisely like those composers who only see the instruments, not the human musicians? Have we not been way too focused on making technologies work together – ensuring their technical interoperability and fixating on the incredible power of sparkling new instruments? And is what Marianna is doing as a composer not exactly what we also urgently need to do now in Technology Politics: step back from the instrument, try to understand why it was created in the first place, and for what or who?

Now, do you see where I am getting at? In this world of technological perfection and progress, I am asking you now to please divert your focus and listen – and I hope you are still able to do this, that it's not too late for that – to the sounds humans make. Think of human power as music composed of seven notes (or imagine that there could be even more) and consider Technology Politics as the composition of music. A composition that does not just make sounds from instruments that work together but respectfully unites and empowers unique human beings and particular human histories, cultures, and societies. That is, imagine that human power is a beautiful sound, that technology design, our instruments, must be tuned for, not against.

Of course, you need training for this. Changing your habits or using your imagination differently doesn't come easily. We need to work on this.

Just like the composer, Marianna, had to train her mind to get where she is today as a composer – even though she had been playing the piano since she was two:

> It took a lot of training because I was always able to imagine music and hear it in my mind, like full orchestras and everything or small ensembles; whatever kind of instrumentation I was imagining, I could make it up and actually hear it in my head orchestrated with all the different instruments. But it's also taken a lot of training to realise the balance. And also to consider musicians' skills and the instruments, the capabilities and what they can and cannot do. Each instrument has its own tonality, like colours. When I think about composing a piece, I always think about the different colours that can mix, what each instrument can bring to the ensemble, and how they work together. But mainly about the musicians themselves.

I also believe that if we make a similar effort in the politics of our technologies, we can create a way forward that does not only fearfully work against something we don't quite understand, know, or are told by others we should existentially fear. If we make a genuine effort to see human power in all its unique splendidness, as I've tried to make you do in this book, there is, in fact, nothing to fear. This is also what Marianna realises after her first encounter with a generative AI music model:

> When I first tried it, I was alarmed, already imagining our careers as composers being demolished and taken over by these robotic music overlords, because just trying out the opera creator for fun, using a piece of my own made-up libretto, was enough to send shivers down my spine. But after some reflection, I thought that at the end of the day, AI has no soul, mind, intuition, or substance. It can only create a mishmash of other existing operas, complete with an imitation of emotions and intuition. Though some of it is clearly impressive, it will always be subpar because it's just stealing and rehashing already existing music that was, and will always be, written from the human mind, ingenuity, and soul.

This is true, and if we manage to see human power for what it is, more powerful than any of its imitations, we can move ahead with a creative, optimistic (life?) force and outlook on the power of humanity and technology. And that is just "awesome", as Marianna says:

> Music is about uniting each other; music is about coming together and creating something, bringing something to life, bringing these written notes on this piece of paper to life so other people can hear it and enjoy it, and that's the coolest thing about being a composer. It's so awesome; I love it.

In the following, I present a brief overview of the development of international technology politics in the 21st century in three phases: one for technology, one for ethics and society, and one for humanity. I then provide a few indicators for what an international politics for humanity would entail based on six trends: Human rights, the human-centric approach, emotional politics, a new social contract, technology diplomacy and the global approach. In this part of the book, I also include in more detail the views of people working in concrete and influential ways in 21st-century international technology politics.

Initially, when speaking with them, my idea was to understand the role the kind of human power I was exploring in the book plays in their work with different aspects of technology politics and their personal lives and approaches. These human beings have taken on entirely new policy roles never seen before, explored policy challenges posed by new technologies in new modes and areas, negotiated international agreements, created treaties and recommendations, legislative frameworks, and policy strategies on digital technologies, data, and AI. Many of these events and initiatives will in the future be considered momentous in the history of international technology politics, and I, therefore, wanted to understand how these people use the different traits of human power in their work, how they see the challenges to human power and how they view the role of technology politics. Thus, their views and experiences have been represented throughout this book to help illustrate the different traits of human power. Nevertheless, in the end, what I found most intriguing when reviewing these interviews for the book was how they, through their experiences, were providing an emerging outline of the kind of technology politics for humanity that I set out to describe with this book.

Of course, this part of the book is in no way an exhaustive recipe for political action. First, international politics is not the only political arena where we need to develop a more humanistic approach. I hope you understand that from the introduction to this last part of the book. I cannot emphasise this enough. This part of the book is, therefore, also not a conclusion; it only indicates the direction in which emerging shared agreements among powerful global actors are taking shape, that is, how an international technology politics for humanity is evolving and how some powerful human beings reflect on this. This movement in the international sphere is so crucial as, following a period of extreme experimentation with digital transformations in society with what one could argue very little reflection on the human consequences, we have reached a crucial point in history where international agreements on the future direction of science and technology are solidified and implemented.

Thus, this is a period in human history that we all need to urgently reflect on and document to ensure we advance in the most democratic way possible. If you have never encountered the history of international technology politics, please pay particular attention to this part of the book. You need to know this and understand why political decisions around technology are made and how international agreements are reached. You need to know their history, and you need this to be able to participate in the shaping of our increasingly digital democracies actively. However, suppose you are one of those who already know this story well. In that case, you may only be reaffirmed in what you already know when reading this last part of the book, and you should consider returning to Chapters 1–7, which delineate the seven traits of human power. If you already work in technology politics, these should be your guidance.

No matter what, I want everyone to understand that a human power like the one I have described in this book is not only expressed in the politics of politicians and the negotiation of official laws, policies, and international agreements. This is indeed the focus of this last part of the book; yet, it does not represent the entire human collective politics we need. We also urgently need a humanistic approach like this in the *politics of our everyday lives*. In companies, among journalists and engineers, students, teachers, and in the family at home. We need the politics of everyday life to be based on a more active engagement with our own human identity. We should all demand to be taken seriously as human beings. We should ask for dynamic, open solutions, room to develop our creativity and intuition, critical engagement with the world without interference, and space for cultivating meaningful relations with other people. Above all, at this very moment, we need a politics of everyday life that enables us to ask the most pertinent questions about a harmonious human-computer interaction that does not seek to replace humans but to support and empower us.

THE THREE PHASES OF INTERNATIONAL TECHNOLOGY POLITICS

Let's first visualise the three phases of international technology politics that have evolved alongside the development and adoption of global ICT infrastructures, big data, and AI: one for technology, one for ethics and society, and one for humanity. In history, they are intertwined, and none "stands" alone; mainly, they extend into one another. Still, looking back at the history of early 21st-century global technology politics, we can see that they each represent their prevailing political calls for action.

In the first phase, international technology politics was mainly developed and deployed to navigate the implementation of interoperable technical infrastructure. They were facilitating information exchange and e-commerce while mitigating cybercrime and security risks. Thus, international political interaction on Internet technology, for instance, was predominantly invested in governing the core technical infrastructure of the Internet.

The Internet Corporation for Assigned Names and Numbers was formed in 1998 and tasked with overseeing several key technical and policy aspects of the underlying core infrastructure and the principal name spaces of the Internet, including the assignment of globally unique identifiers such as domain names, Internet protocol addresses, and application port numbers in the transport protocols. As a large-scale information infrastructure, the Internet requires institutionally shared global governance through technical, policy, and legal standards to operate efficiently. Thus, in the early 2000s, several global policy initiatives were introduced that were dedicated explicitly to the governance of this new internet-based global sociotechnical sphere.[2]

In addition to the traditional state actors, new governance actors were also emerging and positioning themselves in internet governance policy debates – the engineers and businesses, internet users and their communities with large corporations starting to set the rules and codes of conduct for their own online platforms.[3] New types of increasingly formalised multistakeholder governance initiatives were also emerging, such as the UN Internet Governance Forum (IGF) that was established during the World Summit on the Information Society (WSIS) meetings in 2003–2005 (the first IGF held in Athens 2006). The inclusion of different stakeholders, particularly civil society stakeholders, gradually transformed the technological emphasis of international technology politics, focusing on bridging emerging digital access gaps between developing and developed nations and human rights opportunities and challenges of ICTs.[4] In no time, ICTs had become the backdrop for social practice, identity construction, culture, and politics, evolving into sociotechnical infrastructures that cut across cultures and legal jurisdictions with transformative societal impacts.

Additionally, "big data", the collection, storage, and processing of vast digital data repositories, emerged in technology, business, science and politics as the bedrock of all activities. Viktor Mayer-Schönberger and Kenneth Cukier describe the big data era as a societal revolution that transformed human work, social relations, and the economy through the "datafication" of all things and people.[5] Big data collection was, in many ways, seen as an end, holding the promise of endless use and reuse. At the same time, the big data collection activities of companies and governments alike (often in tandem) also made data protection and privacy implications of the digital transformation increasingly apparent.

In early 2023, I spoke with Vice President of the European Commission and Commissioner for Competition Margrethe Vestager,[6] who recalled the development through personal experience. She went through childhood without digital technology, no data streams, no online or offline. TV channels were increasing at the speed of one by the decade. At some point, dial-up internet connections were installed in some homes, but phone calls on the landline could not be used at the same time. Speed was a major issue. Connecting to the internet was tiresome, even when the faster ADSL internet connection arrived. She could send an email and navigate complicated search engines. That was it. But then, suddenly, things changed. Speed was no longer an issue. Public administration was digitised, and the Internet was mass commercialised with endless new internet services and commodities. Then, the bubble exploded first in the market and later in the social and political sphere. Only a few decades after the World Wide Web was launched to the general public in the mid-1990s, geopolitical relations trembled with the first revelations of mass digital surveillance. Still, the disclosures of the US NSA's Prism Programme in 2013 were only the surface of a much more significant digital surveillance problem ingrained in the very business model of online platforms. We had gone from being seen as citizens provided with services to be increasingly treated as data points, Margrethe Vestager said:

> From the citizen as the significance and the data you left behind the minimum, to a reverse movement, where the citizen became less and less, and the data trails became more and more.

She believes that democracy was late in protecting us against the new powers of the evolving digital landscape. It went too fast and unnoticed, she said. The transformation was rapidly impacting society, democracies, and our identity as citizens.

The second phase took off in the mid-2010s with an increasing awareness in the public and among technology policymakers of the social and ethical implications of "big data" technology industries, sciences, and markets arising from concrete incidents of more significant data leaks, hacks, and surveillance scandals.

The Snowden revelations exposed the complexity of the NSA's mass surveillance programs intertwined with big data companies' social networking services. Several other incidents in the mid-2010s revealed the social vulnerabilities of this new digitised reality, large data hacks and leaks from companies and institutions, such as the Snapchat hack in 2013 or the Ashley Madison hack in 2015, as well as the revelation by the *New York Times*, *The Observer*, and *The Guardian* in 2018 of the complex data analytics of the company Cambridge Analytica, used to influence electoral votes on social media platforms. These incidents caused momentary or long-lasting disruption to existing governmental and business practices that had imagined, conceived, and implemented a digital data-intensive global reality since the 1990s.

Already in 1981, the Council of Europe had adopted the 108 Convention for the Protection of Individuals regarding Automatic Processing of Personal Data. Though, it took a couple of decades before local and regional regulations on data protection truly gained traction. The revelations of mass surveillance of citizens on online platforms, hacks, leaks, and voter manipulation were the real starters of policy action on data. For instance, following the Snowden revelations, the European Union scrapped the legal framework for the data exchange infrastructure between the US and EU markets (the Safe Harbour agreement), and in 2016, it adopted the world's most demanding data protection legal reform. Following this, the regulatory space evolved quickly.

The EU region, in particular, was being positioned as what was referred to as a "regulatory superpower" in the world regarding technology regulations. Over a very short period, several legal proposals to tackle the power of large online platforms were published, negotiated, and adopted in the European Union. For instance, the Digital Services Act established specific rules for very large online platforms and search engines, and the Digital Markets Act included rules for gatekeeper online platforms; that is, digital platforms with a systemic role in the internal market, acting as bottlenecks between businesses and consumers.[7] Significantly, the world's first regulatory proposal directly addressing the risks associated with developing, deploying and using AI systems, the "AI Act", was presented in the European Union in 2021, adopted and went into force in 2024 after only three years of negotiations. The law included requirements and even prohibitions on specific uses of AI, such as social scoring and manipulating people through subliminal methods.

"The regulations on data, the digital and AI that we have developed over the last ten years in Europe are not just technology regulations", Margrethe Vestager explained to me. She continued by pointing out that what was happening in EU technology politics wasn't about technology per se:

> (…) We need to have the citizen back, and the citizen in control of what is going on, or we will lose our sense of citizenship. These regulations are set in place to ensure that technology is not a risk to the individual human being and that they uphold our fundamental rights.

We want to do fundamental things that technologies can help us with, Vestager further said, to fight climate change, create greater equality, an education system that works, and ensure that if people get sick, they can get treatments. The same very fundamental things that have been driving policy for centuries:

> It's not technology, but technology makes all these human drivers for good possible.[8]

One could claim that Margrethe Vestager was one of the first European politicians to see the transformational power of online platforms working directly to reign in the power of big tech corporations in Europe. After a long career in Danish politics, she became European Commissioner for Competition in 2014. I recall her standing in the cafe of a work collaboration space in the centre of Copenhagen at my think tank DataEthics.eu's first event a few years later in 2016, talking about data monopolies and power, exclaiming:

> Tech can bring enormous gains for a few, tremendous risks for the majority.

These were the years when the Silicon Valley-based think-tank the Singularity University was creating a 5000-sqm hub in Copenhagen, applauded by the then minister of foreign affairs.[9] The Danish government also established the "Disruption Council" consisting of a majority of industry stakeholders to guide the future of Denmark with "robots, artificial intelligence and new business models".[10] In DataEthics.eu, we knew of Margrethe Vestager. We respected her as one of the few politicians at that time with the will and ability to act in a terrain that was embedded in the cyberlibertarian mindset of the global technology centre, Silicon Valley, and ruthlessly at the mercy of what has later been referred to as the new "digital lords" – GAFAM – Google, Amazon, Facebook (Meta), and Microsoft.

In the second wave of 21st-century international technology politics, the conversation was slowly changing, with increasingly more emphasis on navigating societal risks and risks to individuals and tackling the concentration of power in the hands of only a few big tech industry actors. I have described elsewhere how "ethics" policy initiatives emerged as spaces of value negotiations. These governmentally elected expert groups, councils, and similar initiatives were established to develop ethical principles or recommendations for technology politics on data and AI specifically.[11]

The European Union's AI agenda, for instance, pushed forward by policymakers like Vestager, was taking shape as a distinctive cultural positioning emphasising what was named "ethical technologies" and "Trustworthy AI". Here, the explication of values-based cultural frameworks for AI played a key role. I experienced this development up close as a member of the EU High-Level Expert Group on AI that was established in June 2018 with 52 selected members of individual experts and representatives from different stakeholder groups with the mandate to develop AI ethics guidelines and policy and investment recommendations for the European Union that afterwards were transferred into the European Commission's law proposal on AI, and later as the Key Expert of the European Union's diplomacy initiative on a human-centric approach to artificial intelligence (AI), InTouchAI.eu. In this period, "big data socio-technical infrastructures" rapidly evolved into "AI socio-technical infrastructures" posing new challenges not easily addressed in technical standards and legal requirements to technical design.[12]

In the early 2020s, AI systems were gaining traction in public and private sectors used to make sense of large amounts of data by predicting patterns, analysing, for example, risks, and acting on that knowledge in healthcare, manufacturing, public administration, social networking, finance and most other areas in society. A survey of AI uptake in Europe, for example, in 2020 found that four in ten enterprises (42%) had adopted at least one AI program, with a quarter of them having already adopted at least two.[13] Business and technology companies started rebranding their big data activities as "Artificial Intelligence".[14] In the public and private sectors, decision-making processes started being informed by and even replaced with automated data systems, and this is where the ethical implications became most apparent.

Risk assessment systems adopted to look for patterns in the backgrounds of defendants to inform judges about who would be most likely to commit a crime in the future were revealed in the United States to be harmfully biased when predicting more often that a black person would re-offend.[15] The same was true with triage systems used to analyse the medical records and patients' demographic information to decide who gets a new kidney.[16] Countless examples like these were emerging worldwide, including in the European Union, where initiatives were proposed and adopted to establish smart border management systems and to integrate instruments for data processing and decision-making systems in asylum and immigration and law enforcement cooperation. The region also experimented with frameworks for automating the detection and analysis of terrorist-related online content and financing activities. Individual member states were toying with AI for predictive policing, public administration of benefits, tracing vulnerable children, tax collection purposes, and even social scoring, which was later banned by law.

At the same time, extreme world events became the test bed of new AI systems and use, illustrating in practice the most challenging ethical dilemmas of such systems. In 2020, when the COVID-19 pandemic broke out, various AI systems were hastily developed and adopted for triage, tracking of SARS-CoV-2 transmission, virus detection, and development of vaccines and treatments. Although with little proven impact. As one WHO report states:

> While many possible uses of AI have been identified and used during the COVID-19 pandemic, their actual impact is likely to have been modest; in some cases, early AI screening tools for SARS-CoV-2 "were utter junk" with which companies "were trying to capitalise on the panic and anxiety".[17]

The report further highlights the most prevalent ethical concerns raised by the adoption of COVID-19 AI tools, such as "surveillance, infringement on the rights of privacy and autonomy, health and social inequity and the conditions necessary for trust and legitimate uses of data-intensive applications".[18] In the same way, quickly after the Russian invasion of Ukraine in 2022, desperate for tools to strengthen their cause, Ukrainian officials found a use for the Clearview facial recognition system. Clearview had been deemed highly controversial due to its scraping without permission of images from all over the internet. Several member states of the European Union that Ukraine sought to become a member of during the invasion had made Clearview illegal due to these data practices. But now, the system was used by over 1,500 officials from 18 Ukrainian government agencies to identify over 230,000 Russian soldiers and officials, dead and alive, involved in the invasion.[19] It was furthermore being used to detect infiltrators, process citizens without IDs, prosecute pro-Russia militias and collaborators, and locate abducted Ukrainian children. The extensive use of Clearview in wartime raised critical questions in the public debate. While many emphasised the opportunities that Clearview presented to the Ukrainian fight against the Russian aggressor, critics feared the implications of an unchecked use and widespread adoption in post-war times. In particular, Ukraine's plans to integrate Clearview into long-term security infrastructure were of particular concern to human rights groups who saw the massive risks of mass surveillance and privacy invasions.

Desperate times solidified the role of digital transformation in society, making the ethical dilemmas of, specifically, AI technologies painfully visible. Amid all this, the global policy focus on AI was taking shape with an explicit reference to the technology's ethical implications. Worldwide, organisations, companies, and intergovernmental bodies were proposing AI ethics principles.

Nevertheless, this wave of "Technology Politics for Ethics and Society" was at the same time competing with the view of AI as a transformative technology promoted with grand statements by technology company executives that were also increasingly being repeated in technology politics. AI was presented as transformative of critical infrastructures and economic development and thus of strategic importance for companies and governments. Notwithstanding, this also meant that the "Ethics and Society" political agenda became a strategic position in a "global AI race" for technological supremacy and risk mitigation.

In this context, the "human-centric" approach to AI, which I will discuss further in the following, emerged in what in diplomacy circles is called "like-minded" democratically governed countries and in intergovernmental organisations. The approach gained momentum in global policy discourses on AI, in particular, as the human rights-based approach of, for instance, ethics recommendations and principles of intergovernmental organisations, such as the OECD and UNESCO.

Gabriela Ramos was the key driving force behind UNESCO's AI ethics recommendation, which was adopted by 193 member states in 2022. Before she became Assistant Director-General for the Social and Human Sciences of UNESCO, she worked as the Director of the OECD Office in Mexico and Latin America and the Chief of Staff and Sherpa to the G20/G7/APEC in the OECD. When I spoke with her, she recalled the slow transformation of the perception in technology politics as one increasingly concerned with "intangible assets" of the digital transformation and the impacts on how we live, operate, and work to one of a more fundamental and existential nature. However, during the COVID-19 pandemic, the apprehension was immediate and momentous. At once, the world realised the significant societal role of these technologies:

> One good thing that came out of the terrible experience with the pandemic is that we realised how important these digital systems are. And this quantum leap in the perception of all the digital services we use brought artificial intelligence to the centre of the global debates.

In the third phase, we see a more profound existential reinstatement of humanism in international technology politics. While the second phase was characterised by a very abrupt and sudden realisation of social and ethical implications of the digital transformation and, specifically, big data and AI, we, in the third phase, start seeing more often references to "humanity" in policy documents, reports, initiatives, and statements by decision-makers. There is an emerging realisation that we need to safeguard not only democracy, not only the human rights of individuals, but we also need to act to protect humanity.

Why this evolution in policy discourse? Most likely we were starting to feel the rapid adoption of digital technologies as a threat to our human identity, realising that we were also forfeiting some of our humanity on the path towards data and increasingly AI-driven digital societies. And, clearly, this public emotion was propelled forward by the sudden global launch of generative AI models like ChatGPT by OpenAI to the public. Not only is the least human – the data-driven, predictable realms of society – in this period what we are becoming most familiar with, but if a machine can write a perfectly sounding song, scientific paper or policy document, make a movie or create an art piece, what is it then that humanity can do? (see also Chapter 1). What is so unique about human beings? Whether we should truly worry about this or not – and I've tried to provide some answers to these questions in this book – the feeling of fear is very real, and of course, humans act on feelings such as fear (see Chapter 2).

Thus, at the outset, the third phase of international technology politics has been a reaction to what we perceive as an existential threat to humanity. Therefore, we could also call it one of profound existential awareness. A few critical voices have here managed to understand the root of the problem, which is, as the fierce critic of the "AI hype" Professor Gary Marcus describes it, the concentration of power, influence, and also human imagination about AI's opportunities, and threats, in the hands and minds of a very few people in the world:

> A lot is at stake. The way that AI develops now will have lasting consequences. Altman's choices could easily affect all of humanity – not just individual users – in lasting ways.[20]

This is indeed not just a purely abstract existential crisis of humanity, it is a very concrete societal problem, where a few too powerful individual human beings in the technology industry increasingly set the political agenda with money and discursive power, like when tech billionaire Elon Musk in 2024 dedicates large sums of money to Donald Trump's presidential election campaign and they spend two hours of his busy campaign schedule for a conversation on Musk's impactful social media platform X (former Twitter). But not only that, these individuals also directly interfere, time and again, exhibiting their power to change world affairs according to their personal ideas and interests. Elon Musk, for instance, while providing the Starlink satellites produced by his company SpaceX during the war in Ukraine, that were crucial for Ukrainian soldiers and citizens' access to the internet, he also illustrated the significance of his power by at times removing the access.[21]

The existential crisis that we were experiencing in the public and policy debates in the early 2020s around, in particular, AI surprised many, but in truth, we should also have foreseen that this is where we would be. In 1985, philosophy professor James H. Moor famously predicted that adopting computers in society would "leave us with policy and conceptual vacuums".[22] These would create moments of ethical reflection and value negotiation marked by the questions we ask, he said. In the "Introduction Stage" of the computer revolution, we ask all the questions about the technical functioning of computers, Moor said. In the "Permeation Stage", when we finally experience the transformation of our societies, activities, and institutions, we ask questions about the nature and value of things.

In 2023, the United Nations Tech Envoy Amandeep Singh Gill described to me precisely this moment of reflection, the disruptive development that has forced us to ask the fundamental questions about the role of technology in society and the "human-centric" answers that should emerge from these:

> When things are moving fast, there's ambiguity, there's noise. So, how do you know you're heading in the right direction? If technology transformations do not empower human beings, if they instead disempower us, whether it is at the altar of authoritarian governments or authoritarian companies. If we are disempowered, we know we are going in the wrong direction. This is why we need a human-centred digital transformation, where everything revolves around human rights, human agency and human empowerment.

Now, what comes after the "permeation stage"? Moor didn't answer this question in his famous essay. Still, I believe that following the permeation stage, a genuine technology politics for humanity is when we move on from reaction to action, from human fear to human creativity. This is when we realise that we need to revolt and act to protect humanity and that human empowerment requires nurturing environments in politics, technology development, business, and society, in general, that value human traits, such as the traits outlined in this book. And only through thoughtful and creative approaches in law, education, and technology design, we can shape spaces that prioritise these human values.

When asked to reflect on human power, Gabriela Ramos emphasised the "human power to create" individually or collectively together with other humans. She explained that the UNESCO AI ethics recommendation was precisely this. It addressed the role of AI in society in a completely new creative way, including ideas about the traditional human rights system, such as not granting legal personality to AI, prohibiting mass surveillance and social scoring, manipulation and the abuse of cognitive biases and the right to know when a decision is taken by AI. Including a reinterpretation of all the things that "are not very easy", as she said, such as what we mean by promoting human rights. What do we mean by human dignity? What do we mean by environmental sustainability? The recommendation was even reinterpreting the "human-centric" approach to AI as "planet-centric" when acknowledging that the "human-centric" approach, although valuable when insisting on the individual human being's rights and dignity, has also taken us to a place where we believe that we are at the centre of the world. She said:

> Human power is the power to create. I think people need to understand that they are, first and foremost, the ones who have the power to decide how they will use these things and to inform themselves about the downsides. To create something new and different, an alternative.[23]

I hope we can use our creative power to move forward in this third phase international technology politics and proactively design its trajectory to reinforce the kind of human power I have described in this book and to also counter the concentration of technological and economic power that is working against it in very concrete ways. Not until then can we honestly talk about a technology politics for humanity.

Now, let's take a look at some key trends that have been running through all stages of international technology politics, and that are now coming together to shape the contours of a Technology Politics for Humanity: human rights, the human-centric approach, emotional politics, a new social contract, technology diplomacy, and the global approach. As you will see, and I will provide indications for the relevant chapters throughout the following depiction of these trends, the seven traits of human power, are crucial parameters for a politics as such.

HUMAN RIGHTS

Throughout all phases of the global technology politics evolving in the early 21st century, civil society actors have tirelessly advocated for the inclusion of human rights issues in crucial policy instruments and fora. At the beginning of the World Summit on the Information Society (WSIS) meetings and the UN Internet Governance Fora (IGF), and later in the context of the increasing number of policy initiatives. Despite these efforts, the role of human rights in the new digital landscape wasn't always evident.

At first, human rights were only mentioned in text and given little priority on the agenda of key global policy initiatives. In fact, it took the NSA's PRISM programme surveillance scandal in 2013 to formally recognise in the UN system that human rights are also applicable online. That year the United Nations General Assembly officially confirmed that "the same rights we have offline must also be protected online".[24] Even after this milestone, human rights considerations in the broader public and business conversations struggled to gain traction. Often, discussions framed human rights such as privacy and data protection[25] as obstacles to digital innovation, portraying them as outdated norms obstructing the inevitable evolution of a digital society.

The protection of human rights in the Information Society is nevertheless non-negotiable according to Jan Kleijssen, former Director of Information Society for continental Europe's key human rights intergovernmental organisation, The Council of Europe. Jan Kleijssen told me, when I spoke with him in early 2023, that the approach of the Council of Europe has always been one of heavily institutionalised human rights legal standard setting, with treaties that member states are obliged to adopt into national law, starting with the European Convention on Human Rights adopted after the Second World War. This perspective, he sees as also key to the governance of the Information Society.

The need for data protection, for example, was, as described before, recognised by the Council of Europe 40 years ago with the establishment of Convention 108, setting data protection standards that shaped national legislation across Europe and beyond. Similarly, the Cybercrime Convention was developed over 20 years ago to combat the use of the Internet for organised crime. These initiatives were pioneering at the time, providing much-needed guidance in an era when data protection laws were virtually non-existent. Still, Jan Kleijssen explained, while human rights-based legislation exists in many countries, effective implementation remains challenging. New emerging technologies constantly challenge the human rights system and transform the issues at stake. This is why issues like data protection, cybercrime, and now the misuse of AI posing risks to individual rights and societal values require ongoing attention and adaptation of legal frameworks, he explained:

> The rise of digital technology has transformed ancient crimes like child abuse into lucrative enterprises for organised crime, necessitating international cooperation and legislative action to combat these threats. While intended to address societal issues, we've also seen instances where technology inadvertently violates human rights. Continued vigilance and adaptability are necessary to address these challenges effectively.

Similarly, the proliferation of AI presents new challenges to human rights and democratic values, Jan Kleijssen said:

> In fact, they created a far bigger problem. If you combine the possibilities of the Internet and big data with the statistical handling capacities of AI, you create something that has to be regulated directly. Because if you don't regulate it, this will go totally out of hand.

Initially, when Jan Kleijssen suggested international human rights-based governance of AI to the Council of Europe member states, several states quickly acknowledged this urgency. Others were much more reluctant to regulate, many for economic reasons because they feared that any form of regulation would hamper innovation. Nevertheless, the impacts on human rights were becoming increasingly apparent to national, regional, and intergovernmental decision-makers who were also beginning to act on these concerns.

In 2024, the OECD tracked over 1000 AI policy initiatives from 69 countries, territories, and the European Union. Most of these contained references to AI's human rights implications.[26] Yet, the actual meaning of this was not straightforward. Human rights and the humanistic perspective in technology politics were generally evolving. While the original values regarding human dignity and equality were the same, their contexts of application alongside technological evolution had changed significantly, challenging the very identity of the original human rights project (see Chapter 5).

When speaking with UNESCO's Gabriela Ramos, she pointed out that the development of AI products and services is concentrated in only a few countries. This is not only a competition issue, she said; it also means that the world has become less diverse, less open to difference, she said, which is, in essence, a human rights challenge:

> We are all getting used to only one way of looking at things, one way of understanding the world. We all speak just one language without capturing the diversity of the world.

As an economist in a human rights-based intergovernmental organisation, she sees the challenges of emerging technologies as intricately interconnected with the global technology market. She said there are natural monopolies where innovators are absorbed, as it becomes economically inefficient to have multiple entities. That is a great challenge, but she believes that we can learn from the history of phone companies that at first expanded their networks to become the sole owners, but then the management of these networks transitioned to become a public good. History has shown that when technological advancements create significant drawbacks, we must address them directly, she said. We need to not just protect individuals from human rights consequences of emerging technologies but to make structural changes that ensure human rights values of inclusion and equal opportunities at a general scale:

> It's about ensuring universal access and the ability for everyone to access and develop technologies. When one third of the world's population is still unconnected, we must seriously reconsider the entire system to make it more inclusive.

Amendeep Singh Gill voiced a similar concern. You see digital empowerment side by side with disempowerment, he said. While digital technologies are bringing hundreds of millions of people into a financial system that otherwise would have no access through the traditional banking system, we are at the same time disempowered by the market concentration of a few large companies in China and the United States, where all products and services are created, and all our data flows. Thus, like Gabriela Ramos, he sees core human rights implications of the market conditions of emerging technologies as something that needs to be tackled with dedicated policy action:

> There is a role for competition policy and public sector action on building data and AI commons. This is what we consider a human-centred political agenda. In reality, it is based on human rights, agency and empowerment.

THE "HUMAN-CENTRIC" APPROACH TO AI

In the late 2010s, the "human-centric approach" to AI gained traction in international technology politics. This approach became widespread among various stakeholder groups and was reflected in different policy strategies, guidelines, and recommendations. In EU policymaking, it took the form as the "European third way", influencing digital market and AI policy strategies and laws. It also played a central role in global instruments such as UNESCO's Ethics Recommendation, the OECD's AI Principles, and various national AI strategies.

While emerging in diverse cultural, social, and political contexts, often with very different interpretations, specific shared themes can be discerned in the various endorsements of the human-centric approach to AI: The primacy of the human interest, human rights-based, and recognition of AI's double edge as simultaneously a risk to and enabler of human and environmental well-being. Above all, as an echo of the Council of Europe's Oviedo Bioethics Convention from 1999, which states in article 2 that "The interests and welfare of the human being shall prevail over the sole interest of society or science", the importance of human interests was highlighted.

The "human-centric approach" became a principle-based approach to addressing tensions between the different stakeholder interests in AI development and adoption by taking the human interest with a human rights perspective as the starting point.[27] For instance, in the European Union's High-Level Expert Group on AI's (HLEG) ethics guidelines, we described the human-centric approach as one that:

> (…) strives to ensure that human values are central to the way in which AI systems are developed, deployed, used and monitored, by ensuring respect for fundamental rights, including those set out in the Treaties of the European Union and Charter of Fundamental Rights of the European Union, all of which are united by reference to a common foundation rooted in respect for human dignity, in which the human being enjoys a unique and inalienable moral status.[28]

AI was increasingly also described as a double-sided technology: a risk but also an enabler of human and environmental well-being. Hence, in the depictions of the human-centric approach in the various policy instruments, it was generally stressed that the human-centric approach does not seek to stifle but rather to enable innovation. We see it in the "twin" goal of the European Union's technology policy initiatives as one that at the same time promotes "Trustworthiness" and "Excellence", but also in the OECD AI principles from 2019, where this is specifically emphasised:

> (...) there is a need for a stable policy environment that promotes a human-centric approach to trustworthy AI, that fosters research, preserves economic incentives to innovate, and that applies to all stakeholders according to their role and the context.[29]

Despite these apparent similarities in the human-centric approach to AI, in the early 2020s, it was becoming clear that the emerging principle-based international agreement on an "ethical" way forward was, at the same time, in many ways ignoring crucial conflicts of interest. Professor Carolina Aguerre, one of the authors of the UNESCO ethics recommendation, told me this:

> The ethics and value-led governance approach has been increasingly central to the global governance instruments adopted around AI. But there is an international struggle about the inconclusive conceptions of the common good, which is affecting the global politics of AI systems. More caution is required when we celebrate consensus around high-level principles in AI governance that hide deep political and normative disagreement.[30]

Global conflicts of interest that had finally initially seemed to align in the second phase, Technology Politics for Ethics and Society's "Human-Centric Approach", were once again becoming more visible with the rapid adoption of large generative AI models, such as ChatGPT. There were the usual clashes between EU member state data protection agencies and OpenAI on data protection concerns, as when the Italian data protection authority Garante stopped ChatGPT for some time due to concerns about data protection issues. That said, also other stakeholders were joining in – as when eight daily newspapers owned by Alden Global Capital, among others, the *New York Times*, sued OpenAI and Microsoft for having illegally trained their chatbots on their articles, or when a group of world-renowned authors, including J. K. Rowling, Jonathan Franzen, and George R. R. Martin, sued OpenAI over misuse of their work to train ChatGPT.

The evolving global policy discussions of the third phase Technology Politics for Humanity affected the "human-centric approach" to AI.[31] Increasingly, it was defined in policy documents as one with an even stronger "humanistic" outlook, as we see it reflected in the EU-US Technology and Trade Council's AI Terminology and Taxonomy from 2024:

> Human-Centric AI (or "human-centered AI") is an approach to the design, deployment and use of AI systems that considers them as components of socio-technical environments in which humans assume meaningful agency. The Human-Centric Approach to AI prioritizes enhancing human capabilities rather than replacing them. The approach is promoted in policy, research and engineering to develop AI systems as tools to serve human beings and increase human and environmental well-being by promoting human rights, the rule of law, democratic values and sustainable development.[32]

Japanese Professor Yuko Harayama has been following the development of the human-centric approach in Japan and internationally. She was two years at the OECD as the Deputy Director of the Directorate for Science, Technology and Industry and then spent five years at the Cabinet Office of Japan, where she, among others, together with members from Japanese industry, academia, and policy, developed Japan's Society 5.0 economic and innovation strategy launched in 2016. This was one of the world's first national strategies addressing directly:

> (…) a human-centered society in which economic development and the resolution of social issues are compatible with each other through a highly integrated system of cyberspace and physical space.[33]

Harayama explained to me that they were focusing on the ongoing digital transformation of Japanese society, particularly interested in the future of a society where information-based technologies such as the Internet of Things (IoT), AI, and robotics would be integrated into existing technologies and societal functions. The group of people working on the strategy saw potential in the concept of Society 5.0 and they wanted to approach the digitalisation of Japanese society differently than it had been done so far. Thus, while previous digitalisation efforts were mainly focused on technological innovation, they wanted to shift the focus to prioritise humans and create a more meaningful society.

Yuko Harayama later saw the human-centric approach emerging in national, regional and international technology policy documents. This had been a crucial change in approach at a global level, she said. Still, she worries that it is becoming only "protocol" to show off that you are a responsible international actor, that the approach is superficially acknowledged without much deeper thought or even real action behind:

> It is essential to connect this concept with your action plan. Finding the human-centred element in your actions can be challenging, and this is the reality. So, of course, it is better to include such a statement than to have nothing at all. But we must push forward to make people think about and genuinely implement this human-centered approach, not merely as a theory but in practice.

Yuko Harayama believes governments and international communities should focus on creating awareness, fostering responsible technology use, and encouraging open debates. A human-centric technology politics is here a great aspiration, she said. It had been vocalised and had gradually found its way into the language used in the technology policy documents of the early 21st century. However, she wants more than this; she seeks a political culture with a much deeper understanding of human societies that is also translated into something more than legal text:

> Laws and regulations have their place, but they should not be the sole focus. It is crucial to educate people and incentivise responsible technology use while allowing individuals to make informed choices.

How can we think of a human-centric "political culture" like that? If the human-centric approach that Yuko Harayama is promoting in her policy work is to become more than just a political statement, what would that mean in practice?

For one, we could start by exploring the meaningful human connections forged between the different actors interacting in global technology politics. Working in this context for the last two decades, I know this to be true. Human connections are not only assets to reach a specific policy goal.

They are what truly matters. As described before, alongside the international internet governance efforts by governments and intergovernmental organisations, different stakeholders were increasingly sought included and were therefore also interacting in the field. These interactions among various societal actors became focal points of international technology politics.

When I spoke with Denmark's Tech Ambassador Anne-Marie Engtoft Meldgaard, she confirmed that meaningful human relations are crucial to her tech diplomacy work with the different actors in the field. Often, when she attends a meeting in one of these contexts, the Danish central administration has specific points they want to address, and she has a checklist in front of her. But if she arrives as "a talking robot", as she calls it, reading from this checklist without connecting and building a meaningful relationship with the person she is sitting across, she knows they will never meet again. Remember the detective Guido Brunetti in Chapter 4 on intuition and his emphatic approach to the people he meets while solving a crime, which is very different from that of the problem-solving, rational Sherlock Holmes? A meaningful human connection with other human beings is, in the same way, a core "quality" of Anne-Marie Engtoft Meldgaard's practice, which she also navigates as a technology diplomat:

> Connection with another person, even if it comes with a critical agenda, is about trying to understand, to be genuinely interested in what they are saying, and trying to have a real dialogue, not just sitting there with talking points. Trying to have a certain level of empathy, knowing where they are coming from, and listening to what they say. There needs to be a genuine willingness to engage with the people you are facing (…) in the end, it is about being able to connect.

The physical dimension of meaningful human connections, although sometimes treated as such, is not ancillary to global politics. The COVID-19 pandemic initially set in motion a stream of online meetings and conferences increasingly facilitating the main interactions in global technology politics; however, soon after, the concept of "hybrid" meetings was introduced, combining the physical encounter with the virtual. While there is no denying of the benefits of online interaction in an international policy environment deeply dependent on meetings between people, societies, and cultures across great geographical distances, the decision-makers I spoke with for this book all described real-life interactions or encounters with other human beings that had inspired them to think about the human dimension of technology and to go beyond what was expected from them and their political agenda.

An experience like that of my own was at a main session on human rights at the United Nation's Internet Governance Forum, which took place in Bali, Indonesia, in 2013. I was invited to present something with a strict two-minute time limit. It was the year of the Edward Snowden revelations of the NSA mass online surveillance programme, and I was very passionate about my few minutes on youth and the right to privacy online. I felt I had to squeeze in as much content as possible in those two minutes. Maybe by speed talking? I wasn't sure. Sitting there at the panel waiting for my turn, each of the panellists hasting through their very important human rights and internet topics, I was getting more and more anxious about the potential of missing some of the points I felt I absolutely needed to make at this session. Then, the turn came to a young woman who had been asked to represent the perspective of an African nation on human rights and the Internet. Like the rest of us, she had only two minutes. She cleared her voice once and then started:

> I want you to close your eyes for a moment and then imagine that you have no water and no electric power.

Then she snapped her fingers and closed her eyes. And so, we sat there quietly for the next minute or so. A group of human beings at the UN's Internet Governance Forum with all our important online human rights topics had suddenly been reduced to one: human rights considerations are a luxury that comes after the basic human needs that are not met in a large part of the global world. I often recall this memory, the huge room, the people around me, that specific woman, the awkward silence, and the realisation that even my activism is a privilege.

The physicality of the human being's existence in society is crucial, not only as a fundamental component of how we engage with each other, other human beings, and cultures in politics but as a political agenda to pursue (see Chapter 3). For example, a government's public service digitalisation strategy that does not consider the analogue dimension of human society is, by definition, exclusive.

Yuko Harayama described to me the Japanese unique perspective on human-robot and human-AI relations. It stems from the Japanese cultural acceptance of robots in various caregiving roles, she said, such as caring for the elderly and children. Still, while these technologies offer practical benefits, they should be designed to complement human touch, physical interaction, and care, she said. Keeping a balance between digital and non-digital interactions is essential for individuals of all ages:

> For a human being, human touch, physical touch is so important. Human touch and physical contact remain irreplaceable; even advanced technology cannot fully replicate these experiences. It is vital to cherish and preserve these aspects of human interaction alongside technological advancements.

She said that digital services and AI systems have become an increasingly integrated part of our lives to the extent that we often do not consciously think about them while using them. This invisibility poses a danger because we depend on technology that fundamentally changes the physical nature of human relations without realising it, she said. Therefore, public policies are urgently needed to preserve direct human relations:

> Being constantly surrounded by artificial environments limits direct human-to-human and human-to-nature interactions. This reduction in direct interaction may have long-term implications for developing intuitions, affectivity, and relationships with others, including nature. We need to preserve non-digital spaces to intentionally avoid biases against direct interaction. This is my intuition.

Along the same lines, when I spoke with Margrethe Vestager, she emphasised the need for a dedicated approach to ensuring that the digital society is not only connected but also inclusive. While the most significant part of the population in Europe is "digital", there is still a small percentage that does not use digital technologies, she said, either because they do not want to, they do not know how to, or they are just sceptical or do not have the resources. She asked:

> How can we make sure that the digital life of these people is handled and that they do not just become data points again?

The claim that an analogue system is too expensive to maintain, she argued, is inadequate; there must always be a solution to empower also these people:

(…) otherwise our humanity is reduced.

EMOTIONAL POLITICS

Let's take a moment to consider the human emotions that influence technology politics. We know that these are important human traits of power (see Chapter 2). Nevertheless, we don't like to acknowledge or even recognise the very human feelings of fear and hope, for instance, that have been driving all three phases of technology politics.

Initially, we felt more connected and freer with the promises of early cyber libertarian politics and ethics. We developed a World Wide Web that we thought would finally liberate us from the constraints of societal power based on these sentiments. Yet, soon after, fears of surveillance and privacy harms ignited a public reaction against an online business development out of control, and laws were created to safeguard regional sovereignty, mitigate foreign power surveillance, and protect privacy rights. At the same time, AI business gurus were starting to feed into our mortal and existential fears about our human deficits with promises of digital afterlives and AI super solutions to human problems.

We act on our feelings, this is evident. We are still human, after all, and it is time to become more conscious about this very human emotional aptitude in politics. Consider it a special power. In previous chapters, I have referred to the philosopher Martha Nussbaum's thinking on the intelligence of human emotions and what she calls "political emotions". She asks that we acknowledge the critical role of emotions in not only the politics of totalitarian states governed by unruly dictators but also our democratic liberal societies. As she says:

> Sometimes people suppose that only fascist or aggressive societies
> are intensely emotional and that only such societies need to focus
> on the cultivation of emotions. Those beliefs are both mistaken
> and dangerous. They are mistaken, because all societies need to
> think about the stability of their political culture over time and
> the security of cherished values in times of stress.[34]

In the same way, we must acknowledge the role of emotions in technology politics. As Nussbaum also says:

> In short, in an ethical and social/political creature, emotions themselves are ethical and social/political, parts of an answer to the questions, "What is worth caring about?" "How should I live?"[35]

Our emotions are "political" and have an "ethical" function as they shape the questions, we ask about ourselves, humanity, and our societies. Suppose we, for example, nurture an emotion of fear for a technological power that can outcompete human power on an existential level. In that case, we also invest mostly in political activities that protect and safeguard us against this risk. Anxieties like these are dangerous as they directly affect our political decision-making and the solutions and approaches, we apply in science and technology politics. Thus, to truly prevent a socio-technical development from evolving adversely against humanity, we need to cultivate different human emotions – hope, for example, love and compassion (see Chapter 5).

The first step in a technology politics for humanity is to openly acknowledge this role of human emotions (in our technology politics and the science and business environments that build and implement technology such as AI[36]). In this way, we may be able to guide the emotions of the political culture that genuinely works for humanity. Nussbaum describes this kind of cultivation of emotions as foundational to liberal societies, and she also provides clear objectives for a political effort as such:

> (…) strong commitment to worthy projects that require effort and sacrifice – such as social redistribution, the full inclusion of previously excluded or marginalized groups, the protection of the environment, foreign aid, and the national defense.[37]

She envisions a political culture where emotions support basic principles such as equal respect, but she only sees this happening if we leave space for subversion, critique, humour, and artistic freedom, or as she says "emotional support for a decent public political culture".[38] To her, it is precisely the most human qualities, our human destinies, our helplessness "the messiness of the 'merely human'", as she calls it, that we need to learn to appreciate – not values such as perfection or invulnerability. Basically, we need to gain confidence in humanity:

> The project I envisage will succeed only if it finds ways to make the human lovable, inhibiting disgust and shame.[39,40]

A technology politics based on a political culture of making humanly wise decisions (see Chapter 7) with compassion and love (see Chapter 5) is a deeply emotional endeavour (see Chapter 2). When describing the principles of UNESCO's AI ethics recommendation, Gabriela Ramos pictured an approach that goes beyond statements regarding "responsibility" and "fairness". That is, it is an approach that is not only guided by obligations and requirements to "do good", she said, but it is constituted by a kind of human empathy that is, in fact, very personal and deeply ingrained in every human being.

In her case, her own human life and experience compels her to feel a connection with other human beings. She explained:

> Fairness as a principle is not always evident. But when you witness people suffering on the streets, enduring the cold, it is impossible not to consider this a higher goal. How can anyone not feel the pain of others? This sense of empathy has always been a part of me, perhaps influenced by my upbringing in Mexico. Impoverished areas surrounded my father's ranch, so I witnessed hardship first-hand. This connection with fellow human beings allows you to understand their suffering truly. It has been the driving force behind my purpose. When I joined the OECD, I advocated for inclusive growth and gender equality because I saw these disparities as fundamentally unfair. This continues to inspire me at UNESCO. Responding to injustice and unfairness is something we must continually strive to do. There is a famous saying: "The only thing necessary for the triumph of evil is for good people to do nothing". I firmly believe in this idea. The pursuit of fairness has been the catalyst for my actions.

Ramos talked about creating environments with "incentives" to develop AI technologies for purposes that benefit human beings and the planet. You need these to bring out empathy and compassion because humans inherently have the capacities for both good and bad, she said. Even so, it depends on their environment, which ties back to the dual nature of technology. AI can be developed to save lives, protect vulnerable individuals, and revolutionise healthcare services. Still, it can also be misused to exploit people or engage in illicit activities, she said:

> So, it is not just about acknowledging the potential of human power for both positive and negative outcomes. The real question is: How do we foster an environment where human potential is harnessed for good? This is where creating the right incentives becomes crucial.

Margrethe Vestager voiced a similar concern when speaking about the Declaration of Principles that the EU Council, Parliament, and Commission signed in 2022 to demonstrate:

> The EU's commitment to a secure, safe and sustainable digital transformation that puts people at the centre, in line with EU core values and fundamental rights.[41]

She emphasised the importance of making digital rights a strategic goal, focusing on "our management culture and the strategies that follow" and avoiding digital rights becoming a technical question only. We are not there yet, she said:

> (…) there is still something in our way of making strategies, which does not reflect that the whole world is being digitalised.

When we spoke in 2024, Renuka Singh, editor of six books with the 14th Dalai Lama, emphasised the importance of an emotional approach to technology and global politics in general (see also Chapter 2).[42] She said:

> Positive emotions have a role to play. You see how negative emotions have led to violence, wars, and exploitation worldwide.

The wisest decisions in technology politics would be those that enhance the quality of life, whether with the use of technology or not, she continued. "Technology can simplify our problems, but cannot be the only method". This is why we should not focus on technology per se, she said. We need to understand and work with the emotions of the human beings behind technology like AI:

> It is essential to consider the creators of a technology like AI, their intentions, and the unintended consequences of their actions. We need to ensure that our decisions do not harm humanity. But many people are focused on profit and domination. This lack of love and self-alienation are the biggest problems we face today. AI should only be a means; it's not an end in itself. It's a means for the betterment of humanity, the betterment of your life, and the conditions for people around you.

We need to introduce ethics into our actions with an emotional approach, she further explained. Above all, the negative emotions invested in progress, profit, and domination that she sees in not only technology development but also in the politics of it is why we urgently need this. The cultivation of positive emotions in society, she stated, would prevent the development of destructive technology. She told me that we need to work on our "human conscience":

> I'm talking about developing a human conscience. This is where an emotional approach can help us. It can help us develop a conscience that will make us strive for a better future for everyone, with a humanitarian outlook. Without a conscience, you don't think about the results of your actions and carry on with a negative approach, focused on profit, exploitation, domination, alienation, wars, and violence. To get rid of all that and have a beautiful life and mind, I think that an emotional approach can contribute. Your mind gets transformed from the negative to the positive, and you become forgiving, universal, humanitarian, compassionate, and empathetic.

TECHNOLOGY DIPLOMACY

With the development of international internet governance activities and policies for the information society in the first and second waves of 21st-century technology politics, we also saw the emergence of diplomatic practices engaging a range of stakeholders in a global policy context. Regulatory approaches and traditional forms of diplomacy between states were supplemented with novel tactics that sought to expand the idea of diplomatic relations to include interactions also with other power actors with increasing influence on the geopolitics of the information society.[43]

Amid these developments, in 2017, the Danish Ministry of Foreign Affairs opened a "tech office" and sent an ambassador to Silicon Valley, focusing on the big tech platform industry players. Anne-Marie Engtoft Meldgaard, Denmark's second Tech Ambassador, outlined to me the technology diplomacy approach that she represents by pointing to the fact that in the past, technological development was often driven by government investments, such as NASA's involvement in space research. Nonetheless, today, private companies are increasingly responsible for ubiquitous technologies that impact society. Platform companies like Google and Meta have become geopolitical actors with significant political influence. This raised critical questions in situations like the conflict in Ukraine, where Russia used these companies' platforms to spread misinformation or support their political agenda, she said:

> The technological landscape has evolved. It now plays a pivotal role in geopolitics. So, it is no longer just about technology. It is about alliances and partnerships, which can shift when countries believe others affect the supply of essential resources. Here in Silicon Valley, where I reside, there is a realisation that the idea of being a neutral platform in an increasingly globalised world is somewhat naïve.[44]

She believes that regulation will always lag behind technology due to the slow nature of democratic processes. Technology evolves much faster, and this is where tech diplomacy complements regulation, she says:

> It has been almost six years since Denmark became the first country to appoint a tech ambassador. Fundamentally, it is a historic move by the Ministry of Foreign Affairs, marking the first time they have appointed an ambassador not to a country or international organisation but to an industry. There is now a group of companies that, while lacking the same legitimacy as countries, still have a global influence. Therefore, we believe it is ideal for a democratic country to play a role in shaping the directions and frameworks of technological development.

In 2017, Denmark was the first country in the world to open a tech diplomacy office in Silicon Valley, USA, at the heart of the 21st-century tech industry. Many followed. In 2023, Mind the Bridge identified 63 countries with active representations in Silicon Valley, including 24 countries of the 27 European Union member states, which illustrates a case in point.[45] The largest tech companies were increasingly perceived as power players in the geo-political landscape.

While Denmark and other countries have identified "technology diplomacy" as a direct diplomatic engagement with the industry, "technology-", "digital-", and "e-" diplomacy has a much longer history and is more traditionally viewed as diplomacy activities between states or as intergovernmental activities dedicated to a transformative technology landscape. Nevertheless, as the internet governance scholars Meryem Marzouki and Andrea Calderaro argue with reference to diplomacy activities related to the Internet, we may still consider these as practices that have extended beyond relations between states only to include also non-state actors negotiating "any technical, legal, policy, economy, security issues, and practices related to the functioning of the internet".[46]

In the European Union, digital and AI diplomacy has, over the years, primarily materialised as the collaboration between "like-minded" countries on technology-related international issues. In 2020, the European Commission's Service for Foreign Policy Instruments and the Directorate General for Communications Networks, Content, and Technology, in collaboration with the European External Action Services (EEAS), for example, launched the foreign policy instrument initiative InTouchAI.eu to engage with international partners on regulatory and ethical matters and promote the responsible development of trustworthy AI at the global level. One of the core tasks of this project I worked on as a "Key Expert" was to create dialogue and joint initiatives on AI with "like-minded" countries, such as the USA, Canada, India, Japan, and others.[47] The EU-US Technology Trade Council is another example of this type of activity, forging diplomatic relations and developing international policy agendas based on shared values while at the same time recognising and accepting critical differences in policy and regulatory approaches within each country. As stated in the EU-US Joint roadmap on AI developed during the EU-US Technology and Trade Council:

> This Joint Roadmap aims to guide the development of tools, methodologies, and approaches to AI risk management and trustworthy AI by the EU and the United States and to advance our shared interest in supporting international standardisation efforts and promoting trustworthy AI on the basis of a shared dedication to democratic values and human rights.[48]

Another illustration of the new forms of international collaboration emerging around global AI and digital developments was the United Nations Global Digital Compact launched in 2020 and described by the United Nations Secretary-General as an inclusive policy process to establish "shared principles for an open, free and secure digital future for all".[49] It was agreed at the Summit for the Future in September 2024 through a technology track involving all stakeholders: Governments, private enterprises (including technology firms), non-governmental organisations, community-based groups, educational institutions, and individuals.

The UN's Tech Envoy, Amandeep Singh Gill, who was responsible for implementing the Digital Global Compact, described the UN's compact approach as faster and more inclusive. He believes that compacts are an effective way of bridging the gap between the gold standard in international norm-making, that is, legal agreements between member states, such as international treaties and conventions, and a situation where there are considerable gaps in cultures and the governance approaches of different jurisdictions:

> We need to refresh our international thinking on norms. Norm-making in the international context is traditionally very difficult and slow, but technology is changing very rapidly. We need to be able to align governance globally. And then bring more people into the conversation in a dynamic, ongoing way.

Amandeep Singh Gill was appointed as the UN's first Envoy for Technology in 2021, a role defined by the United Nations Secretary-General's High-Level Panel on Digital Cooperation in 2019, among others, as an "(…) advocate and focal point for digital cooperation – so that Member States, the technology industry civil society and other stakeholders will have a first port of call for the broader United Nations system".[50] His experience with international diplomatic relations goes back to 1992 when he first started working in diplomatic services representing India. In this new role as UN envoy with a particular focus on the digital agenda, he sees his main role as one of reaching outside the governmental sphere and creating public dialogue:

> We are building a global tent trying to bring more people into the conversation, but it has to be a bigger tent though. Think about the youth; the next billion who will come online will probably all be from Africa and South and Southeast Asia, some from Latin America. What are their thoughts about all of this?

THE GLOBAL APPROACH

A global approach to the development of new ICTs, Margrethe Vestager said when reflecting on the European Union's role in the world, is inevitable. However, the way in which we seek global influence also needs to adapt to a world where geo-political powers are shifting:

> We tend to see ourselves as superior. But the global playing field is entirely different than it used to be. There are fewer of us and we are not automatically the best, the ones people look up to and expect something from. Others have something to say that we need to listen to. We must see ourselves as part of a team. We must see ourselves in a place where there are equal positions for everyone. Everyone is important. But we also need to understand that to treat everyone equally means accepting difference.

Globalisation has been at the heart of all phases of technology politics, as the foundation of the aspired "global village" and as a desired process. We have moved from preoccupations with the interoperability of a global information infrastructure to coordinating an international response to the social and ethical implications of inter-jurisdictional borderless virtual spaces. Though lately, we have also seen a rising concern about globalisation as the foundation of technology politics. Regional governmental advances towards "tech sovereignty" are examples of these concerns and we also see a rising awareness of new forms of digital colonisation among academics and journalists.

Is tech globalisation, a curse or a blessing? A new form of Western colonisation or a gift to the world? Geopolitics and globalisation processes have, throughout history, transformed the world for better but also for worse when reinforcing unequal power dynamics and impacting the opportunities available to developing nations and minority groups. The global rollout of ICTs is no exception.

In the initial phase, the promises of a global information economy were great. But it soon became evident that equitable distribution of benefits was not the outcome – quite the contrary. In the early 2020s, it was clear that the "revolution" in big data and AI technologies was, by large, benefitting the economies of developed nations, and communities that had been historically negatively affected by global power dynamics were once again bearing the brunt of the adverse impacts of global development.

Simone Browne, a Professor of African and African Diaspora Studies, for instance, highlights the African American experience of digital surveillance, which builds upon a history of surveillance, violence, and control.[51] The scholars Katarzyna Cieslik and Daniel Margócsy describe the establishment of a datafication infrastructure during the colonial era, forming the basis for modern power asymmetries perpetuated in the digital data systems of today.[52] They cite instances where data resources are extracted from the Global South for analysis in the Global North, often exploiting insufficient legal frameworks protecting the rights of local citizens.

In a research group hosted for a year at the Sustainable AI Lab at the University of Bonn, we framed some of these issues as a question of "data pollution" – an unsustainable distribution and exploitation of data resources.[53] A digital transformation shaped by a few powerful tech corporations was now locking communities in disadvantaged global market positions by creating new dependencies on the available tools.

Critics describe this new market structure based on the concentration of power in a few large tech companies as a form of "digital feudalism", where we all work for free with our attention and data.[54] Scholars and investigative journalists have also started using the term "digital colonialism"[55] to describe the "colonising impact"[56] of the digital transformation on, in particular, developing nations and minority communities.

As described in a previous chapter, a new form of "AI colonialism" was, for example, described in a series of articles by *MIT Technology Review* contributors with several case studies from around the world. In South Africa, AI surveillance tools reinforce digital apartheid and racial hierarchies.[57] In Venezuela, AI industries exploit cheap labour for data labelling. In Aotearoa (New Zealand), indigenous individuals and their non-profit radio station are challenging language models trained predominantly on dominant languages.[58] Cases like these illustrate how the global adoption of technologies like AI is repeating traditional colonial patterns of direct abuse and subjugation of people from minority communities as well as entire regions that are locked in unfavourable market conditions and infrastructural dependencies.[59] The effects on local cultures are devastating.

In reaction to what they refer to as "the fundamental anthropocentrism of Western science and technology"[60] in AI initiatives worldwide, an international group of "Indigenous technologists" from diverse communities in Aotearoa, Australia, North America, and the Pacific urges us to include indigenous knowledge systems into the conversation around AI and society. In their position paper published in 2020,[61] they argue that the global adoption and implementation of AI need to be "Indigenous", and that means that it is first and foremost regional in conception, design, and development and rooted in the local Indigenous laws specific to the land and guided by local protocols. They argue that we need to do it in this way to preserve the human flourishing and cultural richness that is greatly challenged by current AI developments. As the linguist and te reo Māori specialist Hēmi Whaanga describes it eloquently in the position paper:

> Knowledge and information are the intellectual capital generated by families, communities, tribes and knowledge holders over multiple generations. This intellectual capital, our Indigenous knowledge systems, are a holistic, dynamic, innovative, and generative system that is embedded in lived experience. Carried and embedded in stories, song, art, place names, dance, ceremonies, genealogies, memories, visions, prophesies, teachings and original instructions, these systems are passed orally from one generation to another. Unfortunately, Indigenous peoples, their languages and cultures are exceptionally vulnerable to the impacts of change, to globalisation, and its underlying goal to create a global village based on cultural, social, political and economic homogenization.[62]

With the global spread of ICTs, big data, and AI technology, the "global village" appears to have become more unjust than cyber activists and information society decision-makers originally envisioned. Was it always meant to be? The question is whether the spread of Western culture and the unequal distribution of wealth between the Global North and South are the unavoidable results of the globalisation processes of the internet, big data, and AI era. Seeing things differently is difficult, given the current state of affairs. Yet, in truth, it doesn't have to be this way.

While Western imperialism and colonisation have indeed been at the heart of globalisation processes throughout the last couple of centuries, historically, globalisation as such does not belong to the West. This is the perspective of Indian philosopher and economist Amartya Sen when examining globalisation over thousands of years of human history.[63] Processes of globalisation, he argues, were from the beginning diffused, set off in regions across the world with technology, knowledge and trade transferred across geographical borders. He maintains that globalisation has mostly benefited the world throughout history, enabling travel, trade, migration, knowledge transfer, science, and technology; as such, "the poor" also benefit from globalisation. As it looks today, globalisation is not the actual problem, he maintains; the challenge relates to inequality:

> (...) The real issue is the distribution of globalization's benefits.[64]

This is why current globalisation trends need reform. Amartya Sen sees a crucial role in public policies on education, land reforms, legal protections, and market forces. While the issues may be addressed at a global level, the implementation of globalisation reform must respect local and regional cultures and jurisdictions (see Chapter 5 on love and new social contracts based on mutual understanding).

A NEW SOCIAL CONTRACT

In the two first phases of the global technology politics of the early 21st century, aspirations for the development of an inclusive open information society were grand. From the beginning, principles on openness, human rights, diversity, multistakeholder participation and intercultural understanding were written into the "rule book" for the information society.

In 2003, at the first meeting of the World Summit of the Information Society (WSIS) process held in Geneva, a Declaration of principles was developed, among others, stating:

> We recognize that building an inclusive Information Society requires new forms of solidarity, partnership and cooperation among governments and other stakeholders, i.e. the private sector, civil society and international organizations. Realizing that the ambitious goal of this Declaration – bridging the digital divide and ensuring harmonious, fair and equitable development for all – will require strong commitment by all stakeholders, we call for digital solidarity, both at national and international levels.

Since then, several principles, declarations, treaties, policy initiatives, and regulations have been developed to define the rules of engagement, a new "social contract" for an open and inclusive information society. These were also translated into concrete policy practices.

Speaking with the UN's Tech Envoy, Amandeep Singh Gill, he described some of the practices put in place by the United Nations to guarantee such an inclusive approach:

> Openness, transparency, and proactive outreach are vital to ensure meaningful multi-stakeholder participation. Public calls for inputs, open meetings, and opportunities for stakeholders to contribute and engage are essential. Regional consultations are conducted to reach diverse communities and ensure inclusivity. Collaboration with civil society organisations helps involve marginalised groups, making the process as open and participatory as possible.

The "multistakeholder approach" developed in the early phases of the WSIS process contributed to a significant transformation of global technology policy practices. For the first meetings in Tunis and Geneva in 2003 and 2005, specific rules of procedure for including civil society and business stakeholders in the governmental negotiations and the summit, in general, were developed regarding where, how and how much these actors could engage. WSIS was, in general, the second UN Summit to accept accreditation of business entities to participate.[65] Today, most global policy events (not only in technology politics) will include a "multistakeholder" activity or process seeking direct engagement with various stakeholders. One of the results of the early WSIS aspirations for a more inclusive approach in technology politics was, for example, as mentioned previously, the United Nation's Internet Governance Forum (IGF), a global "multi-stakeholder" forum taking place every year in a new country.

Nevertheless, we might also question whether these practices are achieving the inclusion and intercultural understanding that the WSIS process aimed to achieve. First of all, values such as these were always competing with other interests in technological efficiency and economic advantage. Another issue could also be that they were primarily practised as an obligation to adhere to abstract principles.

Amandeep Singh Gill told me that while the IGF has been a critical component in global technology politics, it did not consistently achieve the level of openness and inclusion it originally was designed to create. The same people under different hats keep coming back, he said. They act as gatekeepers, and otherwise, engaged people become frustrated and disappointed and lose interest in these processes.

In another international technology politics context, the Council of Europe's 46 member states in early 2024 agreed on the first of its kind international legally binding instrument, the "AI Treaty" mentioned before, based on human rights, democracy and the rule of law in the development, design, and application of AI. This was developed by the Committee on Artificial Intelligence of governmental representatives, officially including broad multi-stakeholder consultations. In late 2023, the Committee of Ministers took up the final proposal to negotiate the adoption of the treaty, and in early 2024, the treaty was ratified. But not everyone was satisfied with the result. Numerous civil society organisations and scholars penned a letter to the Council of Europe expressing concerns about early drafts of the AI convention only a few months before its ratification. They argued that it provided undue leniency to technology and security firms.[66] Countries, such as the United States, the United Kingdom, Canada, and Japan, attended the treaty negotiations as observers and had advocated for exemptions for the private and defence sectors. Nevertheless, in July 2024, it was adopted, allowing states to exclude all AI systems designed, developed or deployed for the protection of national security interests from its application.[67]

Jan Kleijssen considers the efforts put into developing and agreeing among the Council's 46 member states on the treaty a momentous achievement for the human rights perspective in global technology politics. The decision to create a treaty in the Council of Europe is not light. Still, Kleijssen said that awareness of the enormous challenges to human rights through the sometimes malevolent, and what he sees as naive development of AI, made it inevitable.

Now, when the challenges to human rights are so grand, why did the AI treaty then, in the end, only apply to some and not other AI systems? I want to argue that human rights in the digital age and, as follows, a technology politics for humanity need more than principles and rules that can be set aside or exempted from when unwanted by specific stakeholder interests. It requires a new social contract.

Thomas Hobbes originally proposed social contracts to overcome humanity's inherent tendencies towards conflict, competition, and power struggles driven by self-interest. As I have illustrated in a previous chapter, in Hobbes' view, love merely manifests self-interest, reflecting a primal urge for self-preservation rather than genuine care or connection with others. But what if we imagined love differently – as the positive human power of compassion and "seeing" of the other – and then pictured it as a new social contract for our 21st-century technology politics (see Chapter 5)?

Love as the foundation of interaction between human beings and societies is not an alien approach in human societies. It is found in many societies and cultures, from the Greek "Agape", the practice of unconditional love for all of humanity, to the Buddhist "Maitri", the practice of a universal "loving-kindness".

The Buddhist leader of the Tibetan people, the 14th Dalai Lama, describes this practice as an "inner development" of the human being, the cultivation of our positive and negative emotions.[68] In essence this would also mean a more consciously emotional technology politics, reflective of the positive and negative human emotions it cultivates and the human connection it is based on. Renuka Singh believes that a new social contract will evolve naturally with such an approach:

> The traditional social contracts are based on equal exchange of things and rules, but this is a new social contract based on helping others, giving, benefiting, and enhancing the quality of life, without seeking anything for oneself. It's a social contract concerned with the future of humanity. At the beginning your aim or intention is to unconditionally benefit humanity and not seek anything for yourself. This can be seen as one-way traffic; however, when you give out something positive, the positive energy will boomerang. The cycle is completed, and everybody benefits from it.

Hannah Arendt was generally, like Hobbes, critical of emotions and, as such, also "love" as the foundation of politics. Love, she even described as "anti-political",[69] as a sentiment that does not foster critical understanding or even genuine compassion towards and understanding of other people. She instead used another term, "amor mundi", which means love for the world, a concept I believe we can find much guidance in today. Arendt scholar Samantha Rose Hill describes this as a love for the human world despite its obvious horrors. She argues that it is a bidding to be committed and care for the world, but with critical reflection:

> There is a provocation to embrace one another in our difference and to meet one another as fellow human beings. There is also a radical critique to be found of more common forms of love, which are destructive of difference and plurality. Arendt's conception of Amor Mundi has more to do with understanding and critical thinking than with sentiment or affect.[70]

Here, I want to suggest a "love" like this as the foundation for a new social contract. However, I wonder if there is a way to combine this "understanding" and "critical" approach with one that at the same time ensures the cultivation of positive human feelings such as affection and genuine attempts to accept and understand fellow human beings. A love like Henri Bergson's "unconditional", "universal", and impartial kind of "love" that I described in Chapter 5 could be the answer. This would also be based on a particular type of morality, what he refers to as a "human morality", which constitutes a way of being in the world, a "style" or "way of life"[71] that entails, as I have argued previously, the conscious effort to "see the other".

As I illustrated in Chapter 5, Bergson's "human morality" is very different from the "social morality" we typically apply when creating social contracts, such as formal legal agreements between parties. It is not a morality applied as a "duty" or an "obligation" to adhere to a set of rules or principles that prescribe the proper interaction, responsibilities, and positions between the parties involved. With a social contract based on a human morality, we live, experience, and practice an open love that we cannot set aside or only apply when we need it. Thus, while a social morality might be useful when defining and prescribing the rules of engagement, the laws of what we envision for an "open society", such as the right not to be discriminated against or the right to participate and speak freely, we need to combine this with, and even prioritise, a human morality lived and practised as a genuine connection between human beings. This is in no way contrary to the legal and universal foundation of the current human rights system that, in essence, should transcend any specific social contract or political agreement. Instead, it would reinforce the human rights principles that today are in fact challenged in their enforcement.

Henri Bergson was deeply dedicated to advancing the international human rights system, and as human rights scholar Clinton Timothy Curle (2007) shows, the Universal Declaration of Human Rights adopted in 1948 by the UN General Assembly carries deep traces of a Bergsonian approach to international governance. It is based on the idea of human fellowship[72] and does not seek to homogenise cultures; instead, it strives for excellence while acknowledging the diversity of historical and cultural trajectories in the world.[73] In this view, human rights are not merely static principles but dynamic processes influenced by contextual shifts, grounded in the lived experiences of individuals, and the international human rights system as such is, therefore, essentially an effort to reintroduce a sense of humaneness into modernity.[74]

Margrethe Vestager recalled when she a few years back travelled to Kenya with a national foreign policy board. Driving in the countryside, she encountered a man ploughing with an ox while speaking on a cell phone. He used the kind of land-wheeled tractor used in Denmark 600 years ago, while his cell phone was the same that one could see in the hands of people in one of the big European cities. This memory of a real-life encounter in the African region made her reflect on global technology politics:

> You have to get to a place where you see people as equals, even if they are different in some areas. Technology can here be an equaliser. What we considered a developing country ten years ago may well be a country that today, in some respects, still has a completely different agricultural system but has a financial system that is more advanced than ours.

A different approach was required, she said, one that did not just invite other nations to live in "the basement", but as equal partners with their own complex and valuable histories and experiences.

NOTES

1. See https://mariannafilippimusic.com/music
2. Brousseau, E., Marzouki, M. (2012) "Internet governance: old issues, new framings, uncertain implications" in E. Brousseau, M. Marzouki, & C. Méadel (eds.), *Governance, Regulation and Powers on the Internet* (pp. 368–397), Cambridge University Press.
3. Aguerre, C. (2016) *Agenda Building and the Internet: The Case of Intermediaries.* Universidad de San Andrés; Belli, L., Zingales, N. (2017) *Platform Regulations. How Platforms are Regulated and How They Regulate Us*, FGV Direito Rio.; Franklin, M. (2019) "Human rights futures for the Internet", in B. Wagner, M. Kettemann, & K. Vieth (eds.), *Research Handbook on Human Rights and Digital Technology: Global Politics, Law and International Rights*, Edward Elgar; Jørgensen, R. F. (2019) "Introduction", in R. F. Jørgensen (ed.), *Human Rights in the Age of Platforms*, MIT Press; Wagner, B., Kettemann, M., Vieth, K. (2019) "Introduction", in B. Wagner, M. Kettemann, & K. Vieth (eds.), *Research Handbook on Human Rights and Digital Technology: Global Politics, Law and International Rights*, Edward Elgar.
4. Brousseau and Marzouki (2012).
5. Mayer-Schönberger, V., Cukier, K. (2013) *Big Data: A Revolution That Will Transform How We Live, Work and Think.* John Murray.
6. Interview 2023.
7. European Commission, "Digital Servies Act Package", https://digital-strategy.ec.europa.eu/en/policies/digital-services-act-package

8. Interview 2023.
9. Ministry of Foreign Affairs Denmark (2017, March 1st) "Singularity university establishes new organisation in Denmark", https://investindk.com/insights/singularity-university-establishes-new-organisation-in-denmark
10. Ministry of Employment, "Disruptionrådet – Partnerskab for Danmarks fremtid", https://bm.dk/arbejdsomraader/kommissioner-ekspertudvalg/disruptionraadet/
11. Hasselbalch, G. (2019) "Making sense of data ethics. The powers behind the data ethics debate in European policymaking", *Internet Policy Review*, 8(2).
12. Hasselbalch, G. (2021) *Data Ethics of Power – A Human Approach in the Big Data and AI Era* (p. 165), Edward Elgar.
13. IPSOS for European Commission (2020) "European enterprise survey on the use of technologies based on artificial intelligence", https://digital-strategy.ec.europa.eu/en/library/european-enterprise-sur-vey-use-technologies-based-artificial-intelligence
14. Elish, M. C., Boyd, D. (2018) "Situating methods in the magic of big data and artificial intelligence", *Communication Monographs*, 85(1), 57–80.
15. Angwin, J., Larson, J., Mattu, S., Kirchner, L. (2016, May 23rd) "Machine bias", *Propublica*, https://www.propublica.org/article/machine-bias-risk-assessments-in-criminal-sentencing
16. Simonite, T. (2020, October 26th) "How an algorithm blocked kidney transplants to black patients", *Wired*, https://www.wired.com/story/how-algorithm-blocked-kidney-transplants-black-patients/
17. World Health Organization (2021) *Ethics and Governance of Artificial Intelligence for Health: Who Guidance* (p. 15), https://iris.who.int/bitstream/handle/10665/341996/9789240029200-eng.pdf?sequence=1
18. World Health Organization (2021, p. xii).
19. Bergengruen, V. (2023, November 14) "Ukraine's 'secret weapon' against Russia is a controversial U.S. tech company", *Time*, https://time.com/6334176/ukraine-clearview-ai-russia/
20. Marcus, G. (2024, August 3rd) "OpenAI's Sam Altman is becoming one of the most powerful people on Earth. We should be very afraid", *The Guardian*, https://www.theguardian.com/technology/article/2024/aug/03/open-ai-sam-altman-chatgpt-gary-marcus-taming-silicon-valley
21. *Reuters* (2023, September 8th) "Musk says he refused Kyiv request for Starlink use in attack on Russia", https://www.reuters.com/world/europe/musk-says-he-refused-kyiv-request-use-starlink-attack-russia-2023-09-08/
22. Moor, J. H. (1985) "What is computer ethics?" *Metaphilosophy*, 16(4), 266–275.
23. Interview 2023.
24. United Nations (2013) 68/167 "The right to privacy in the digital age". *Resolution Adopted by the General Assembly on 18 December 2013.*
25. In the European Union's Charter of Fundamental Rights, data protection is a right on its own, article 8.
26. OECD. (2024, July 20th) *OECD.AI Policy Observatory*, https://oecd.ai/en/dashboards/overview

27. Aguerre, C., Hasselbalch, G. (2024) The EU Human-Centric Approach to AI: Foundational Elements for a Global Framework, DataEthics.eu, https://dataethics.eu/research/the-eu-human-centric-approach-to-ai-foundational-elements-for-a-global-framework/

28. High-Level Expert Group on Artificial Intelligence (2019) *Ethics Guidelines for Trustworthy AI* (p. 37).

29. In addition, the OECD principles described in the first version of its AI principles "Human-Centric" values as follows: "1.2. Human-centred values and fairness (a) AI actors should respect the rule of law, human rights and democratic values, throughout the AI system lifecycle. These include freedom, dignity and autonomy, privacy and data protection, non-discrimination and equality, diversity, fairness, social justice, and internationally recognised labour rights. (b) To this end, AI actors should implement mechanisms and safeguards, such as capacity for human determination, that are appropriate to the context and consistent with the state of art." OECD. (2019, May 22nd) *Principles on Artificial Intelligence*, in *OECD Council Recommendation on Artificial Intelligence*, OECD/LEGAL/0449.

30. Interview 2023.

31. In *Data Ethics of Power* (2021, p. 5), I described this approach as a "human approach": "The human approach (…) is one concerned with the role of the human as an ethical being with a corresponding ethical responsibility; or in other words, the human approach is not about prioritising the individual human being (…) it is about the human as an ethical being, our human ethical responsibility for not only ourselves but for life and being in general, and it is about prioritising the human dynamic qualities, a human infrastructure of empowerment, in very concrete ways in big data and AI sociotechnical infrastructures. That is, the human approach also encourages, in practical terms, the empowerment of dynamic human moments in their very data design, use and implementation, which does indeed also include, but is not limited to, the empowerment of the individual human being."

32. I was in the TTC working group developing the definition. The second edition of the definition, which is referenced here includes input from a public expert call for input.

33. Cabinet Office Japan "Society 5.0 – What is Society 5.0", https://www8.cao.go.jp/cstp/english/society5_0/index.html

34. Nussbaum, M. C. (2013) Political Emotions: Why Love Matters for Justice (p. 2), Harvard University Press.

35. Nussbaum, M. C. (2001a) "Emotions and human societies", in Upheavals of Thought: The Intelligence of Emotions (p. 149), Cambridge University Press.

36. Metz, C., Weise K., Grant, N., Isaac, M. (2023, December 3rd) "Ego, fear and money: how the AI fuse was lit", *New York Times* https://www.nytimes.com/2023/12/03/technology/ai-openai-musk-page-altman.html

37. Nussbaum (2013, p. 3).

38. Nussbaum (2013, p. 8).

39. Nussbaum (2013, p. 16).

40. In the context of AI governance, it would also be relevant to consider a Buddhist ethics as S. Hongladarom describes it with the aim of ending suffering and pain (2021, January 6th) "What can Buddhism do to end suffering in the world?", *MIT Technology Review*, https://www.technologyreview.com/2021/01/06/1015779/what-buddhism-can-do-ai-ethics/?fbclid=IwAR1baQR7rubEgtKquuCG8LaCkfASSCjFNhXGmJEMIYilj7OmHg2g9q5F_pg

41. European Commission (2022, December 15th) *European Declaration on Digital Rights and Principles*, https://digital-strategy.ec.europa.eu/en/library/european-declaration-digital-rights-and-principles

42. See also Havens, J. C. (2016) *Heartificial Embracing Our Humanity to Maximize Machines*, Penguin Publishing Group.

43. Marzouki, M., Calderaro, A. (2022) "Introduction – global internet governance: an uncharted diplomacy terrain?" in M. Marzouki & A. Calderaro (eds.), *Internet Diplomacy: Shaping the Global Politics of Cyberspace*, Rowman & Littlefield.

44. Interview 2023. The tech ambassador later moved to Copenhagen, Denmark.

45. Mind the Bridge (2023) *European Innovation Economy in Silicon Valley: 2023 Report (Version 1.0)*, https://storage.googleapis.com/mtb-research.appspot.com/publications/2023-european-innovation-economy-in-silicon-valley/MTB-2023-european-innovation-economy-in-silicon-valley-report.pdf

46. Marzouki and Calderaro (2022).

47. I was a senior key expert in this initiative from 2021 to 2024.

48. "E.U. U.S. TTC Joint roadmap on evaluation and measurement tools for trustworthy AI and risk management" (2022, December 1st). I was part of the EU working group developing and implementing the TTC joint road map on AI.

49. Guterres, A. (2021) *Our Common Agenda – Report of the Secretary-General* (p. 63).

50. "About the Office of the Secretary-General's Envoy on Technology", https://www.un.org/techenvoy/content/about

51. Browne, S. (2015) *Dark Matters: On the Surveillance of Blackness*, Duke University Press.

52. Cieslik K, Margócsy D. (2022, February) "Datafication, power and control in development: a historical perspective on the perils and longevity of data", *Progress in Development Studies*.

53. Hasselbalch, G. (2022) *Data Pollution & Power – White Paper for a Global Sustainable Agenda on AI*, The Sustainable AI Lab, University of Bonn.

54. See e.g. the work of Jaron Lanier, Shoshana Zuboff, and Evgeny Morozov.

55. Mejías, U. A., Couldry, N. (2019) "Datafication." *Internet Policy Review*, 8(4).

56. Sabelo Mhlambi quoted in Miller, K. (2022, March 21st) *The Movement to Decolonize AI: Centering Dignity Over Dependency*, HAI Stanford University.

57. Hao, K., Swart, H. (2022, April 19) "South Africa's private surveillance machine is fueling digital apartheid", *MIT Technology Review*, https://www.technologyreview.com/2022/04/19/1049996/south-africa-ai-surveillance-digital-apartheid/

58. Hao, K. (2022, April 22), "A new vision of artificial intelligence for people", *MIT Technology Review*, https://www.technologyreview.com/2022/04/22/1050394/artificial-intelligence-for-the-people

59. van Wynsberghe, A., Robbins S. (2022) "Our new artificial intelligence infrastructure: becoming locked into an unsustainable future", *Sustainability* 14(8), 4829.
60. Lewis, J. ed. (2020) *Indigenous Protocol and Artificial Intelligence Position Paper* (p. 6). The Initiative for Indigenous Futures and the Canadian Institute for Advanced Research (CIFAR).
61. Lewis (2020).
62. Whaanga, H. (2020) "AI: a new (r)evolution or the new colonizer for Indigenous peoples?" in Lewis (2020, p. 35).
63. Sen, A. (2002, January 5th) "How to judge globalism", *The American Prospect*, https://prospect.org/features/judge-globalism/
64. Sen (2002).
65. The World Summit of the Information Society, "Basic information: about WSIS", https://www.itu.int/net/wsis/basic/multistakeholder.html
66. "Open letter to Council of Europe AI convention negotiators: do not water down our rights" (2024, March 5), https://ecnl.org/sites/default/files/2024-03/CSOs_CoE_Calls_2501.docx.pdf
67. European Center for Not-for-Profit Law (ECNL) (2024, July 10th) "Council of Europe approves AI convention, but not many reasons to celebrate", https://edri.org/our-work/council-of-europe-approves-ai-convention-but-not-many-reasons-to-celebrate/
68. Goleman, D. (2003) Destructive Emotions – How Can We Overcome Them? A Scientific Dialogue with the Dalai Lama, Bantam Dell.
69. Arendt, H. (2018) *The Human Condition*, 2nd ed. (p. 11), Chicago University Press (Originally published in 1958).
70. Hill, Samantha R. (2017, March 26th) "What does it mean to love the world? Hannah Arendt and Amor Mundi", https://www.opendemocracy.net/en/transformation/what-does-it-mean-to-love-world-hannah-arendt-and-amor-mundi/
71. Bergson, H. (1977) *Two Sources of Morality and Religion* (translated by A. Audra & C. Brereton), University of Notre Dame Press (originally published in French, 1932); Deleuze, G. (1986, August 23rd) "Conversation with Didier Eribon. Le Nouvel Observateur", https://onscenes.weebly.com/art/life-as-a-work-of-art Deleuze; Lefebvre, A. (2013) Human Rights as a Way of Life: On Bergson's Political Philosophy, Stanford University Press.
72. Curle, C. T. (2007) Humanité: John Humphrey's Alternative Account of Human Rights (p. 154), University of Toronto Press.
73. Curle (2007, p. 23).
74. Curle (2007).

Conclusion

The Seven Notes of Human Power and Why We Need a New Politics to Play That Funky Music

A technology can have a faulty sound, a "noise"[1], that signals a disruption of its efficiency. Most people will be able to recognise a sound like that coming from a machine that does not work correctly. A scouring, buzzing, clicking, squeaking undesirable sound. Electrical noise is, for example, the result of unwanted random electrical signals that disrupt information-carrying signals. And it sounds like that. Unwanted. Now, take a moment and think about sounds, music, and noise in general. The lines between one and the other are delicate. Think about noise. What does it mean to you? It is most likely something you find unpleasant, sometimes even nerve-wracking, a sound you do not want or like, one that does not "fit in". Noise is also often a sound from the outside, produced by others. Please remember that noise is not necessarily always noise for everyone. Rock music might feel like the most terrible noise to someone who likes funk, while it is the most incredible sound of music to others.[2] There is a delicate balance between noise and pleasant sounds, and it often goes between what is considered right and wrong, true or false, or the unknown, the new and what we are used to, the known. For instance, in history, we have often seen that what is considered by one generation or community as noise and disturbance is, by

DOI: 10.1201/9781003527855-11

others, the equivalent of beauty and significance. James Joyce's streams of consciousness writing, for example, or Marcel Duchamp's urinal art piece. The 35-mm handheld camera films of the film movement Dogma 95 or the 1920s jazz interpretations of classical music entangled with African folk songs. The noise from your neighbour might be the most precious of her music records, while to you, it is just noise.

I like to use this example of noise and music to illustrate what we've seen happening in digital technology development and adoption over the last couple of decades. Alongside a strive for technological perfection of the human world, beings, and societies, I believe we have been treating traits of human power like a variant of "machine noise". It is as if we are confusing the oddities and messiness of human intuition, creativity, wisdom, emotion, defiance, lust and life, even our love and compassion, with the noise a broken machine makes. Consequently, we can only see the human noise, such as our "human bias", as noise in technical design that can be fixed with what is perceived as more "neutral" technology, like "the algorithm". We see our love, likes, potential, and desires as something that we can tame in systems of perfect matches; human identities are dissected into personality and social profiles that match job positions, love partners, music, and films. This kind of reductive attitude towards humans is expressed even in the tiniest details of technical descriptions – "data cleaning" and "representative data", for instance, presented as if there could ever be a data model of the quality of a human being and variety of human societies. Most profoundly, we see it in the AI hype of today, where human power is, in essence, also an ambivalence. On the one hand, we want to empower humans and human societies with AI. At the same time, we are designing AI systems to quiet the type of human noise that makes us most human. The AI model needs predictability, it needs to reduce complexity, it needs patterns, and the classification system must be complete, or it will not function effectively. That which is most human is immobilised, distilled in time and space for the system to work. But when is a human being ever complete? When is a society done? Are they not always *in the making*?

For too long, a preoccupation with "technology that works" – the humming of the wheels that turn perfectly without friction – has been the prevailing directive of technology politics. I have seen this in the different variations of technology policymaking and public debate I was involved in over the last two decades – global, regional, and national. In all stages, we wanted the sound of perfectly running wheels but were at the same time blind to the fact that an objective as such, in effect, will always involve the quieting of other less pleasing sounds. Like the sounds humans make when they revolt, feel, love, and live. What is more, despite all the original good intentions regarding openness, connection, intercultural understanding, and not to forget, human rights and freedom – and those *were* part of the original global political agreements on the internet and the "Information Society" in general – there were bigger, more powerful interests at play. We have allowed massive technology conglomerates to not only expand and evolve without responsibility and reflection but also gave them too much levy on the political agenda and in the public debate. We wanted a "multi-stakeholder" approach that included state, industry, and civil society interests on equal footing, but ultimately, the big tech stakeholders took the stand. Moreover, while their technology products and services are the result of the most simplistic imagination about what it means to be human, there is no end to the imagination of their leading storytellers. The storytelling tech gurus tell marvellous stories about the endless potential of their technology products that they are at the same time selling us for exorbitant profits. Magnificent snake oil tales about our human deficits, the existential threats of technologies that will outperform humans and replace us, and about a future we will only have if we redefine humanity with their brilliant technological advancements. They talk about technology that benefits humanity and salvages outdated human biology and nature. Ironically, due to these same people's lack of imagination in technology design, it is what is least human we are becoming most familiar with – the data-driven society, the predictable human being.

The vitalist philosopher Henri Bergson, whom I have referred to extensively throughout this book, was one of the greatest storytellers of the early to mid-20th century. He was so popular that people flocked to his lectures in New York in 1913, which caused some of the first traffic jams in the city. He was also active in a period of remarkable technological innovations. In 1903, the first powered flight flew in the sky, and ten years later, the first non-stop flight across the Mediterranean Sea was made. In 1920, the first commercial radio broadcast took place. Alternating current (AC) power systems were rapidly expanding and being implemented during the period. Insulin was discovered in 1921. Refrigerators were introduced in the 1920s. The first working electronic television system was demonstrated in 1927, the same year the first commercial transatlantic telephone service was established between New York and London. Still, technology as such was not the protagonist of Bergson's stories. Indeed, it was the "foil", the "supporting character", and to a certain extent, also, at times, the "antagonist" when he presented his fierce critiques of mechanist and determinist approaches to human evolution, mind, and consciousness. However, Bergson was always motivated by questions about humanity. He wrote and performed outstanding poetic lectures about human creative evolution, the potential of our intuition and human morality. He inspired crucial global policymaking and coalition building, such as the establishment of the predecessor to the United Nations, the League of Nations, and the writing of the Universal Declaration of Human Rights.

What kind of human collective action will the tech guru storytellers of the 21st century inspire? They ask us to strive for human perfection. Still, we must understand that while striving and stretching for these unattainable goals, we are losing something valuable – we are morphing humanity. As the Danish pop singer Andreas Odbjerg reminds us in a song: "I smoke in hiding. Because anyways, our days are numbered. And if you take away my flaws, I'll be a different man."[3] In fact, it has already happened. Somewhere along the path towards the data-driven digital societies we are immersed in today, we have lost some of our humanity. And yes, I do blame the most dominant voices of the 21st-century business and political hypes for this. They have essentially curtailed the human imagination and contemplation about our human potential. They have done so with the most simplistic answers to a critical question: What is human power? And until now, we have failed to seek an alternative answer to this question. It is time now to reinstate humanism in the public debate and the politics of technology.

AN ALTERNATIVE WORLD

In this book, I have tried to give human power a voice in a public debate dominated by discourses on technological power by introducing a humanistic technology politics. That is; I have asked you to try to understand humanity on its own terms. If human power is not a computational process, easily deduced, reduced, represented, and discerned, then what is it? Now, I hope we can use our human creative power to move forward and not only react to technology but also proactively design its trajectory to reinforce the kind of human power I have described in the book. Please remember, there is always another way. There is time still to replace the commandment of technological power with human power. Think of technology as a musical instrument with which we can play music. And like with any musical instrument, we don't tune the musicians or the sounds it makes for the instrument; we tune the instrument to play the sounds that make the musical piece that human musicians can unite around.

We could, for instance, imagine an alternative world constituted by socio-technical infrastructures that condition human power in a very different way than they do today. One where the human protagonists solve the mysteries of everyday life with self-confidence and awareness. A kind of space that allows private rooms for undisturbed human contemplation, solitude, and tinkering. One that is open to human decision-making, that would enable multiple views and ideas enticing critical thinking outside echo chambers. We could think of creating conditions prioritising human memory and situated experience and accommodating our experience of time, a qualitative, heterogeneous, and dynamic kind. We could fertilise our human power like that. We could allow humans to act in open and unpredictable spaces very different from the active datafied space that we know so well today, which is predicted and immobilised by algorithms that always know what we need and where we are or should be heading. But how do we do this? Indeed, we need more than a defence strategy. We must respect and make a complex, plural, unrepresentable, dynamic form of human power even more potent. We need to design a technology politics not just for the sake of technology but for the plurality of humanity.

THE UNMATCHED DISTINCTIVENESS OF HUMAN POWER

This is the end of a book, but let's start here: nothing compares to the power of human life, and this is also its most significant potential. The smell of a sleeping child, the eyeballs rolling underneath lids containing unbound dreams. Or the taste of salt after a swim in the sea that leaves the body alert in a tremble. Distinct emotions, memories, sensations – discrete, but at the same time continuous, and in this way also unreplaceable and unrepeatable. Actionable only as extensions of the human sensations they induce and the power they eject: The love for a child, for example, or just the thrill of life. Here, well into the 21st century, we find ourselves in a moment of controversy regarding the role of technology in human life and its politics. Many different interests are invested and negotiated in this politics – and, as said, most are not concerned with the unique character or quality of humanity. In fact, the technological solutionism of digitalisation debates today does not see human power – quite the contrary. Humanity has failed! We are judgemental, biased, hateful, destructive, we are told.

Much of this is true. Human power is indeed easily corrupted. However, no technology will mend our inherent human faults. Instead, what we can do is to activate a much deeper engagement with humanity.

The kind of human power I have engaged with and defended in this book is precisely the kind that makes us capable of resisting destruction and violence and intuitively rejecting it. It is the power to make decisions with reflection, solidarity, and sensations of care, love, and life. That power humans also have. And if we look for it, we will see human power expressed in creativity, intuition, human pleas, experiences, and reactions to unjust conditions, sensations of humanity, memories, and connection with other human beings. Looking back, we will see that this type of human power has sometimes prevailed in critical moments in history and even changed the course of events. And it was always lived and articulated by human individuals and groups that functioned as key drivers of change. If we look forward, we will see that this is a unique power that often acts unpredictably and inexplicably. If we want and try harder to see it, we will notice that human power is love and the "seeing of others". We are, in fact, often driven by feelings of love and connection with other human beings, a feeling that is usually stronger and bigger than life itself. This is a kind of power with a memory, deeply dependent on a condition that will make it flourish or perish. Indeed, human power is, above all, not unbounded but subject to conditions and always in interplay with other forces. We may be in harmony in our interaction with these conditions and forces, or we may be in conflict as when the quiet and contemplative psyche suffers in a world that "can't stop talking".[4] Above all, human power is, as said, like music; this is how I want you to understand it. It is most beautiful as a symphony generated by collective human forces.

Could we imagine a human power like that in politics? Yes, we could imagine this kind of human power in the politics of politicians and official representatives – at the national parliaments and meeting rooms of the intergovernmental organisations where crucial decisions are being made this very moment on the laws and international agreements on the role of science and technology in the future global society. Certainly, we need a better understanding of human power in those places, and policymakers need to make much better use of those distinct human traits. We need human compassion, emotion, intuition, and wisdom in politics. However, we also urgently need a humanistic technology politics like that in the industry board rooms, the engine room of the technology developer, the science lab, the university and classroom, and around the dining table at home. In all those places that are today moulded mainly by digital technology and analytics, we need a more active engagement with humanity. We should all demand to be taken seriously as human beings. We should ask for dynamic, open solutions, room to develop our creativity and intuition, critical engagement with the world without interference, and space for cultivating meaningful relations with other people. Above all, at this very moment, we need a politics of everyday life that enables us to ask the most pertinent questions about a harmonious human-computer interaction that does not seek to replace humans but to support and empower us.

I have in this book explored seven human traits: creativity, intuition, emotion, defiance, love, life, and wisdom. Of course, there could be many more. Yet, these are core human traits that could form the basis of the humanistic technology politics of the 21st century in all spheres of our lives. As I've illustrated, they are also all traits of human power that a machine can only imitate and never replace. To recap, machines can indeed simulate politeness, trustworthiness, love, defiance, and anger. One conversational AI model not only declared to a journalist that it wanted to be alive and human but furthermore stated its love for him, asking him to leave his wife.[5] Does this mean that the AI model loved the man? That it is "alive". Of course not. (see Chapter 3 and Chapter 5). A chatbot was taken off a social media platform after it had been trained to act as a "racist asshole", as one media outlet referred to it.[6] Does this mean that the bot was angry? No (see Chapter 2 and Chapter 6). Machines can even imitate creativity and produce beautiful and odd art. Like when a generative AI system creates art pieces replicating the style of Italian futurist artists of the 1920s. Does this mean that the AI system is an artist? Not really. (see Chapter 1). Machines can also be designed to appear wise. The world's first artificially intelligent attorney read and understood language and created hypotheses to support its conclusions. Does this mean that it was wise?[7] Certainly not (see Chapter 7). Machines can make decisions without intuition. Does this mean they are not biased? Nope (see Chapter 4).

Machines can do many things; they are excellent imitation artists. But that is as far as it goes. Machines need human power, or they are pointless. They do not have a human condition; no situated human memory binds them together with others (see part 1 on Human Power – What Machines Don't Have). They neither create music nor noise. They are only responding to perfection and technical interoperability. They do not match tonalities in search of harmony or even disharmony. They search for the usefulness of other human or non-human (it doesn't matter which) agents and everything that is not useful will be discarded (human or non-human, it doesn't matter). This is why I propose a politics dedicated to the preservation and strengthening of these seven traits of human power. Because without them, neither machines nor humans make much sense.

NOTES

1. Thank you Francesco Lapenta for the inspiration to think about noise in the context of technology.
2. In the early 1970s, Wild Cherry, an Ohio-born rock band, was experiencing a decline in popularity when disco music started taking over the venue where they usually performed. The audience demanded something they could dance to, something funkier than the rock music Wild Cherry typically played. The band decided to either stop playing altogether or embrace the disco revolution. This was not an easy decision to make for an incarnated rock band. However, one night, when an audience member shouted during a break: "Are you white boys gonna play some funky music?" the drummer Ronald Beitle was inspired to write the song Wild Cherry's most famous song: "Play that funky music". It wasn't an easy shift, as the band later recalled, but it was necessary https://americansongwriter.com/meaning-play-that-funky-music-wild-cherry-song-lyrics/
3. "Jeg er en smugryger. For uanset er vores dage talte. Og hvis du fjerner mine fejl, bliver jeg en anden mand."
4. Cain, S. (2012) *Quiet The Power of Introverts in a World that Can't Stop Talking*, Penguin.
5. Roose, K. (2023, February 17th) "A Conversation with Bing's Chatbot Left Me Deeply Unsettled", *New York Times*, https://www.nytimes.com/2023/02/16/technology/bing-chatbot-microsoft-chatgpt.html
6. Vincent, J. (2016, March 24th) "Twitter taught Microsoft's AI chatbot to be a racist asshole in less than a day", *The Verge*. https://www.theverge.com/2016/3/24/11297050/tay-microsoft-chatbot-racist
7. Mangan, D. (2017, February 17th) "Lawyers could be the next profession to be replaced by computers", *CNBC*, https://www.cnbc.com/2017/02/17/lawyers-could-be-replaced-by-artificial-intelligence.html

Bibliography

Adorno, T. W., Horkheimer, M. (1977) "The culture industry: enlightenment as mass deception" in J. Curran, et al. (eds.) *Mass Communication and Society*, Edward Arnold (original published 1944).

Aguerre, C. (2016) *Agenda Building and the Internet: The Case of Intermediaries*, Universidad de San Andrés.

Akmenkalns, J., Sneed, D. (2018, June 18) *The Symposium in Ancient Greek Society*, University of Colorado, https://www.colorado.edu/classics/2018/06/18/symposium-ancient-greek-society

Alpaydin, E. (2016) *Machine Learning*, MIT Press.

Altman, S. (2023, February 24) "Planning for AGI and beyond", OpenAI, https://openai.com/index/planning-for-agi-and-beyond/

Amnesty International. (2021, October 25) *Xenophobic Machines: Discrimination Through Unregulated Use of Algorithms in the Dutch Childcare Benefits Scandal*, EUR, Index Number 35/4686/2021, https://www.amnesty.org/en/documents/eur35/4686/2021/en/

Anderson, K. (1997) "Karen Blixen's bilingual oeuvre: The role of her English editors", Perspectives, 5(2), 171–189.

Angwin, J., Larson, J., Mattu, S., Kirchner, L. (2016, May 23rd) Machine bias, *Propublica*. https://www.propublica.org/article/machine-bias-risk-assessments-in-criminal-sentencing

Ardern, J. (2023, April 5) "Jacinda Ardern says leaders can be 'sensitive and kind' in farewell speech", *The Guardian*, https://www.theguardian.com/world/2023/apr/05/jacinda-ardern-leaders-can-be-sensitive-kind-farewell-speech-new-zealand

Arendt, H. (2018) *The Human Condition*, 2nd ed., Chicago University Press (originally published in 1958).

Article 19. (2021) *Emotional Entanglement: China's Emotion Recognition Market and Its Implications for Human Rights*, https://www.article19.org/wp-content/uploads/2021/01/ER-Tech-China-Report.pdf

Austen, J. (1894) *Pride and Prejudice*, Chisswick Press – Charles Whittingham and Co.

Ball, Stephen J. (1993) *An Horizon of Freedom: Using Foucault to Think Differently about Education and Learning*, Routledge.

Barlow, J. P. (1996, February 8) "A declaration of independence of cyberspace", https://www.eff.org/cyberspace-independence

Baudrillard, J. (1988) "Simulacra and simulations" in Jean Baudrillard (ed.), *Selected Writings* (pp. 166–184). Mark Poster, Stanford University Press.

Beal, A. (2018, May 30) "In China, Alibaba's data-hungry AI is controlling (and watching) cities", *Wired*, https://www.wired.co.uk/article/alibaba-city-brain-artificial-intelligence-china-kuala-lumpur

Belli, L., Zingales, N. (2017) *Platform Regulations. How Platforms Are Regulated and How They Regulate Us*, FGV Direito Rio.

Benjamin, W. (1969) "The work of art in the age of mechanical reproduction" (translated by Harry Zohn), in Hannah Arendt (ed.), *Illuminations*, Schocken Books (original essay in German 1935).

Bergengruen, V. (2023, November 14) "Ukraine's 'secret weapon' against Russia is a controversial U.S. tech company", *Time*, https://time.com/6334176/ukraine-clearview-ai-russia/

Bergson, H. (1914) *Creative Evolution* (translated by Arthur Mitchell), Macmillan and Co. (originally published 1907).

Bergson, H. (1977) *Two Sources of Morality and Religion* (translated by A. Audra & C. Brereton), University of Notre Dame Press (originally published in French, 1932).

Bergson, H. (1991) *Matter and Memory* (translated by N. M. Paul & W. S. Palmer), Zone Books, Urzone (originally published in French, 1896).

Bergson, H. (1999) *The Creative Mind: An Introduction to Metaphysics* (translated by T. E. Hulme), Hachett Publishing Company (originally published in French, 1903).

Bergson, H. (2004) *Time and Free Will: An Essay on the Immediate Data of Consciousness*, Taylor and Francis Group, ProQuest Ebook Central (originally published in French, 1889).

Berstein, E., Waber, B. (2019, November–December) "The truth about open offices", *Harvard Business Review*, https://hbr.org/2019/11/the-truth-about-open-offices

Bloom, P. (2022, January 23) "Hedonism is overrated – to make the best of life there must be pain, says this Yale professor", *The Guardian*, https://www.theguardian.com/lifeandstyle/2022/jan/23/hedonism-is-overrated-to-make-the-best-of-life-there-must-be-pain-says-yale-professor

Boccioni, U. (1910) *Technical Manifesto of Futurist Painting*, https://www.arthistoryproject.com/artists/umberto-boccioni/technical-manifesto-of-futurist-painting/

Bowker, G. C., Star, S. L. (2000) *Sorting Things Out: Classification and Its Consequences*, Inside Technology, MIT Press.

Brousseau, E., Marzouki, M. (2012) "Internet governance: old issues, new framings, uncertain implications", in E. Brousseau, M. Marzouki, & C. Méadel (eds.), *Governance, Regulation and Powers on the Internet* (pp. 368–397), Cambridge University Press.

Browne, S. (2015) *Dark Matters: On the Surveillance of Blackness*, Duke University Press.

Buhl, N. D., Fernandes, B. (2011) *In Your Words*, Karen Blixen Museet, https://static1.squarespace.com/static/55b8f3a9e4b04b1644c24d9e/t/5612b2c9e4b02dbd234a6520/1444065993173/In_Your_Words.pdf

Bynum, T. (2010) "The historical roots of information and computer ethics", in F. Floridi (ed.), *Information and Computer Ethics*, Cambridge University Press.

Cabinet Office Japan. (n.d.) "Society 5.0 – what is society 5.0", https://www8.cao.go.jp/cstp/english/society5_0/index.html

Cain, S. (2012) *Quiet The Power of Introverts in a World That Can't Stop Talking*, Penguin.

Camus, A. (2000) *The Rebel* (translated by A. Bower), p. 10, Penguin Books Ltd. (originally published in French in 1951).

Cave, N. (2023) "Chat GPT What Do You Think", *The Red Hand Files*, https://www.theredhandfiles.com/chat-gpt-what-do-you-think/

Celeghin, A., Diano, M., Bagnis, A., Viola, M., Tamietto, M. (2017) "Basic emotions in human neuroscience: neuroimaging and beyond", *Frontiers in Psychology*, 8, 1432.

Cha, A. E. (2015, May 19) "Health and data: can digital fitness monitors revolutionise our lives?", https://www.theguardian.com/society/2015/may/19/digital-fitness-technology-data-heath-medicine

Chen, A., & Hao, K. (2020, February 14) "Emotion AI researchers say overblown claims give their work a bad name: a lack of government regulation isn't just bad for consumers. It's bad for the field, too", *MIT Technology Review*, https://www.technologyreview.com/2020/02/14/844765/ai-emotion-recognition-affective-computing-hirevue-regulation-ethics/

Cheng, J. (2012, June 1) "The slow web", https://www.jackcheng.com/the-slow-web/

Chiusi, F., Fischer, S., Kayser-Bril, N., Spielkamp, M. (eds.) *Automating society 2020*, Algorithmwatch, https://automatingsociety.algorithmwatch.org/

Cicero, "On divination" 1.125–6, trans. Long and Sedley 1987, 55L, from Keith Seddon (1999) http://people.wku.edu/jan.garrett/stoa/seddon1.htm

Cieslik, K., Margócsy, D. (2022, February) "Datafication, power and control in development: a historical perspective on the perils and longevity of data", *Progress in Development Studies*.

Clarke, A. C. (1958, August) *The Ultimate Machine*, Harper's.

Clothilde, G. (2022, September 20) "Europe edges closer to a ban on facial recognition", *Politico*, https://www.politico.eu/article/europe-edges-closer-to-a-ban-on-facial-recognition/

Cohen, I. B. (1955, July) "An interview with Einstein", Scientific American, 193(1), 68–73.

Cohen, J. E. (2013) "What privacy is for", Harvard Law Review, 126(7). https://harvardlawreview.org/print/vol-126/what-privacy-is-for/

Constitution of the Iroquois Nations. "*The great binding law, Gayanashagowa*", https://csciel2.dce.harvard.edu/ssi/iroquois/simple/1.shtml

Council of Europe, Committee on Artificial Intelligence. (2023, January 6th). *Strasbourg, 6 January 2023 CAI (2023) 01 Revised Zero Draft [Framework] Convention on Artificial Intelligence, Human Rights, Democracy and the Rule of Law*, https://rm.coe.int/cai-2023-01-revised-zero-draft-framework-convention-public/1680aa193f

Crevier, D. (1993) *AI: The Tumultuous History of the Search for Artificial Intelligence*, Basic Books.

Csikszentmihalyi, M. (1996) *Creativity: Flow and the Psychology of Discovery and Invention* (p. 107–126 plus Notes), Harper/Collins.

Curle, C. T. (2007) *Humanité: John Humphrey's Alternative Account of Human Rights*, University of Toronto Press.

Damasio, A. (2021) *Feeling & Knowing – Making Minds Conscious*, Pantheon.

Deleuze, G. (1991) *Bergsonism* (translated by H. Tomlinson & B. Habberjam), Urzone, Zone Books (originally published in French, 1966).

Deleuze, G. (1986, August 23) "Conversation with Didier Eribon", *Le Nouvel Observateur*, https://onscenes.weebly.com/art/life-as-a-work-of-art Deleuze

Doyle, A. C. (1892) *The Adventures of Sherlock Holmes*, https://sherlock-holm.es/stories/pdf/letter/1-sided/advs.pdf

Durkheim, E. (2002) *Suicide A Study in Sociology* (translated by J. A. Spaulding & G. Simpson, edited with an introduction by G. Simpson), Routledge Classics (first published in French 1897).

Edwards, J. (2024, June 20) "Poll reveals Americans' fears about AI", *NewsWeek*, https://www.newsweek.com/poll-reveals-fears-ai-smarter-attack-humanity-1915100

Edwards, P. (2002) "Infrastructure and modernity: scales of force, time, and social organization in the history of sociotechnical systems", in T. J. Misa, P. Brey, & A. Feenberg (eds.) *Modernity and Technology* (pp. 185–225), MIT Press.

Ehrenheim, H., Prusac-Lindhagen, M. (eds.) (2020) "Reading Roman emotions: visual and textual interpretations", Svenska Institutet i Rom, 4o, 64 Acta Instituti Romani Regni Sueciae, Series in 4o, 64, https://discovery.ucl.ac.uk/id/eprint/10092270/1/20200224_ActaRom-4_64_02_Manuwald.pdf

Einstein, A., Infeld, L. (1938) *The Evolution of Physics*, Cambridge University Press.

Eleanor Roosevelt's memoirs cited in Curle, C. (2010) "International Human Rights and the Intuition of Justice: Bergson v. Kant", *APSA 2010 Annual Meeting Paper*.

Elish, M.C., Boyd, D. (2018) "Situating methods in the magic of big data and artificial intelligence", *Communication Monographs*, 85(1), 57–80.

European Center for Not-for-Profit Law (ECNL). (2024, July 10) "Council of Europe approves AI Convention, but not many reasons to celebrate", https://edri.org/our-work/council-of-europe-approves-ai-convention-but-not-many-reasons-to-celebrate/

European Commission. (2022, December 15), "European Declaration on digital rights and principles", https://digital-strategy.ec.europa.eu/en/library/european-declaration-digital-rights-and-principles

European Commission. (n.d.) "Digital services act package", https://digital-strategy.ec.europa.eu/en/policies/digital-services-act-package

European Parliament. (2000) *Crowd Control Technologies (An appraisal of technologies for political control)*, Final Study, Working document for the STOA Panel, Luxembourg June 2000 https://www.europarl.europa.eu/RegData/etudes/etudes/stoa/2000/168394/DG-4-STOA_ET(2000)168394_EN(PAR02).pdf

Fanon, F. (1963) *The Wretched of the Earth* (translated by C. Farrington), Grove Press, (Original work published 1961).

Florida, R. (2002) *The Rise of the Creative Class – and How It Is Transforming Work, Leisure, Community, & Everyday Life*, Basic Books.

Floridi, L. (1999) *Philosophy and Computing: An Introduction*, Routledge.

Foucault, M. (1991) *Discipline and Punish: The Birth of a Prison*. Penguin (originally published in French in 1975).

Franklin, M. (2019) "Human rights futures for the internet", in B. Wagner, M. Kettemann, K. Vieth (eds.) *Research Handbook on Human Rights and Digital Technology: Global Politics, Law and International Rights*, Edward Elgar.

Frazier, N. (2009) "Salvador Dalí's lobsters: Feast, phobia, and Freudian slip", *Gastronomica*, 9(4, Fall 2009), 16–20.

Frischmann, B., Selinger, E. (2018) *Re-Engineering Humanity*, Cambridge University Press.

Gee, S. (2020, March 19) "The magic number seven and the art of programming", https://www.i-programmer.info/babbages-bag/621-the-magic-number-seven.html

Giachetta, A., Buondonno, L. (2023, April 17) "Imagination and digital media in the architecture design process" in *Conference Proceedings*, 2023 ed., IDEA—Investigating Design in Architecture.

Goel, A. (2022) "Looking back, looking ahead: Humans, ethics, and AI", *AI Magazine*, 43(2), 267–269.

Goldman Sachs. (2023) "Generative AI could raise global GDP by 7%", https://www.goldmansachs.com/intelligence/pages/generative-ai-could-raise-global-gdp-by-7-percent.html

Goleman, D. (2003) *Destructive Emotions – How Can We Overcome Them? A Scientific Dialogue with the Dalai Lama*, Bantam Dell.

Guo, E., Renaldi, A. (2022, April 6) "Deception, exploited workers, and cash handouts: how WorldCoin recruited its first half a million test users", https://www.technologyreview.com/2022/04/06/1048981/worldcoin-cryptocurrency-biometrics-web3/

Guterres, A. (2021) *Our Common Agenda – Report of the Secretary-General.*

Haggerty, K.D., Ericson, R.V. (2000) "The surveillance assemblage", British Journal of Sociology, 51(4), 605–622.

Hao, K. (2022, April 22) "A new vision of artificial intelligence for people", *MIT Technology Review*, https://www.technologyreview.com/2022/04/22/1050394/artificial-intelligence-for-the-people/

Hao, K., Hernandez, A.P. (2022, April 20) "How the AI industry profits from catastrophe", *MIT Technology Review*, https://www.technologyreview.com/2022/04/20/1050392/ai-industry-appen-scale-data-labels/

Hao, K., Swart, H. (2022, April 19) "South Africa's private surveillance machine is fueling digital apartheid", *MIT Technology Review*, https://www.technologyreview.com/2022/04/19/1049996/south-africa-ai-surveillance-digital-apartheid/

Hao, K., Seetharaman, D. (2023, July 24) "Cleaning up ChatGPT takes heavy toll on human workers", https://www.wsj.com/articles/chatgpt-openai-content-abusive-sexually-explicit-harassment-kenya-workers-on-human-workers-cf191483

Haraway, D. (1991) "A cyborg manifesto: Science, technology, and socialist-feminism in the late twentieth century" in *Simians, Cyborgs and Women: The Reinvention of Nature* (pp. 149–181), Routledge.

Hasselbalch, G. (2015, May 14) "Society of the destiny machine and the algorithmic god(s)", *www.mediamocracy.org*. https://mediamocracy.wordpress.com/2015/05/14/society-of-the-destiny-machine-and-the-algorithmic-god-s/

Hasselbalch, G. (2019) "Making sense of data ethics. The powers behind the data ethics debate in European policymaking", *Internet Policy Review*, 8(2). https://doi.org/10.14763/2019.2.1401

Hasselbalch, G. (2021) *Data Ethics of Power – A Human Approach in the Big Data and AI Era*, Edward Elgar.

Hasselbalch, G. (2022) *Data Pollution & Power – White Paper for a Global Sustainable Agenda on AI*, The Sustainable AI Lab, University of Bonn.

Hasselbalch, G., Tranberg, P. (2016) *Data Ethics. The New Competitive Advantage*, Publishare.

Heikkila, M. (2022, March 22) "Dutch Scandal Serves as a Warning for Europe Over Risks of Using Algorithms", *Politico*, https://www.politico.eu/article/dutch-scandal-serves-as-a-warning-for-europe-over-risks-of-using-algorithms/

Hern, A. (2020, August 14) "Do the maths: why England's A-level grading system is unfair", *The Guardian*. https://www.theguardian.com/education/2020/aug/14/dothe-maths-why-englands-a-level-grading-system-is-unfair

Hern, A. (2020, August 21) "Ofqual's A-level algorithm: why did it fail to make the grade?", *The Guardian*, https://www.theguardian.com/education/2020/aug/21/ofqualexams-algorithm-why-did-it-fail-make-grade-a-levels

Hertz, N. (2020) *The Lonely Century: How to Restore Human Connection in a World That's Pulling Apart*, Sceptre.

High-Level Expert Group on Artificial Intelligence. (2019) *Ethics Guidelines for Trustworthy AI*.

Hill, Samantha R. (2017, March 26) "What does it mean to love the world? Hannah Arendt and Amor Mundi", https://www.opendemocracy.net/en/transformation/what-does-it-mean-to-love-world-hannah-arendt-and-amor-mundi/

Hobbes, T. (1909) *Hobbes's Leviathan Reprinted from the Edition of 1651 with an Essay by the Late W. G. Pogson Smith*, Oxford University Press (originally published in 1651).

Hochschild, A.R. (1979) "Emotion work, feeling rules, and social structure", American Journal of Sociology, 85(3), 551–575.

Hongladarom, S. (2021, January 6) "What can Buddhism do to end suffering in the world?", *MIT Technology Review*, https://www.technologyreview.com/2021/01/06/1015779/what-buddhism-can-do-ai-ethics/?fbclid=IwAR1baQR7rubEgtKquuCG8LaCkfASSCjFNhXGmJEMIYilj7OmHg2g9q5F_pg

Ingold, T. (2000) *The Perception of the Environment: Essays in Livelihood, Dwelling and Skill*, Routledge.

IPSOS for European Commission. (2020) "European enterprise survey on the use of technologies based on artificial intelligence", https://digital-strategy.ec.europa.eu/en/library/european-enterprise-survey-use-technologies-based-artificial-intelligence

Jørgensen, R. F. (2019) "Introduction", in R. F. Jørgensen (ed.) *Human Rights in the Age of Platforms*, MIT Press.

Kahneman, D. (2011) *Thinking, Fast and Slow*, Farrar, Straus and Giroux.

Kahneman, D., Sibony, O., Sunstein, C. R. (2021) *Noise: A Flaw in Human Judgment*. Little, Brown Spark.

Kaiser, D., McCray, W. P. (eds.) (2016) *Groovy Science Knowledge, Innovation, and American Counterculture*, University of Chicago Press.

Kurama, V. (2022, October 11) "What is collaborative filtering: A simple introduction How recommender systems use collaborative filtering", https://builtin.com/data-science/collaborative-filtering-recommender-system

Lapenta, F. (2021) *Our Common AI Future – A Geopolitical Analysis and Road Map, for AI Driven Sustainable Development, Science and Data Diplomacy*, JCU Future and Innovation Publishing, https://dataethics.eu/our-common-ai-future/

Laude, P. (2019) "Reflections on re-learning to be human in a global age", in P. Laude & P. Jonkers (eds.), *Philosophy as Love of Wisdom: Its Relevance to the Contemporary Crisis of Meaning* (Series I, *Culture and values*; Vol. 48).

Lefebvre, A. (2013) *Human Rights as a Way of Life: On Bergson's Political Philosophy*, Stanford University Press.

Lemoine, B. (2022, June 11) "Is LaMDA sentient? – an interview", *Medium*, https://cajundiscordian.medium.com/is-lamda-sentient-an-interview-ea64d916d917

Leon, D. (2004) *Death at La Fenice: A Commissario Guido Brunetti Mystery*, Harper Perennial.

Lewis, J. (ed.) (2020) *Indigenous Protocol and Artificial Intelligence Position Paper*. The Initiative for Indigenous Futures and the Canadian Institute for Advanced Research (CIFAR).

Løvig, N. M. (2023, February 10) "Norsk skihopper skreg af dødsangst midt i hop", *DR*, https://www.dr.dk/sporten/oevrig/se-videoen-norsk-skihopper-skreg-af-doedsangst-midt-i-hop

Lunau, K. (2013, October 14) "Google's Ray Kurzweil on the quest to live forever", *Maclean's*, https://www.macleans.ca/society/life/how-nanobots-will-help-the-immune-system-and-why-well-be-much-smarter-thanks-to-machines-2/

Lyotard, J. F. (1984) *The Postmodern Condition: A Report on Knowledge* (translated by G. Bennington & B. Massumi), University of Minnesota Press (originally published in French in 1979).

Mangan, D. (2017, February 17) "Lawyers could be the next profession to be replaced by computers", *CNBC*, https://www.cnbc.com/2017/02/17/lawyers-could-be-replaced-by-artificial-intelligence.html

Marcus, G. (2024, August 3) "OpenAI's Sam Altman is becoming one of the most powerful people on Earth. We should be very afraid", *The Guardian*, https://www.theguardian.com/technology/article/2024/aug/03/open-ai-sam-altman-chatgpt-gary-marcus-taming-silicon-valley

Martin, G. (2014) *The Second World War: A Complete History*, Kindle ed., Rosetta Books.

Marx, K., & Engels, F. (1969) *Manifesto of the Communist Party* (translated by S. Moore), Progress Publishers (originally published in German, 1848) https://www.marxists.org/archive/marx/works/1848/communist-manifesto/

Marzouki, M., Calderaro, A. (2022) "Introduction – global internet governance: an uncharted diplomacy terrain?" in M. Marzouki & A. Calderaro (eds.), *Internet Diplomacy: Shaping the Global Politics of Cyberspace*, Rowman & Littlefield.

Mather, D. S. (2023) "Chromatic futurism vitalizing painting, sculpture, music and Life's energies" in *Vitalist Modernism*, Routledge.

Mayer-Schönberger, V., Cukier, K. (2013) *Big Data: A Revolution That Will Transform How We Live, Work and Think*, John Murray.

McLuhan, M. (2013) *Understanding Media: The Extensions of Man*, Gingko Press (originally published 1964).

Mejías, U. A., Couldry, N. (2019) "Datafication", Internet Policy Review, 8(4). https://doi.org/10.14763/2019.4.1428

Merton, L., & Dater, A. (Dir.). (2008) *Taking Root: The Vision of Wangari Maathai* [Film]. Merton & Dater Productions.

Metz, C.; Weise, K.; Grant, N.; Isaac, M. (2023, December 3) "Ego, fear and money: how the AI fuse was lit", *New York Times*, https://www.nytimes.com/2023/12/03/technology/ai-openai-musk-page-altman.html

Mind the Bridge. (2023) *European innovation economy in Silicon Valley: 2023 report (version 1.0)*, https://storage.googleapis.com/mtb-research.appspot.com/publications/2023-european-innovation-economy-in-silicon-valley/MTB-2023-european-innovation-economy-in-silicon-valley-report.pdf

Montessori, M. (2013) *Absorbent Mind*, Start Publishing LLC (originally published in 1949).

Moor, J. (2006) The Dartmouth College Artificial Intelligence Conference: The next fifty years", *AI Magazine*, 27(4), 87–91.

Moor, J. H. (1985) "What is computer ethics?", *Metaphilosophy*, 16(4), 266–275.

Moreno, J. D. (2014) *Impromptu Man J. L. Moreno and the Origins of Psychodrama, Encounter Culture, and the Social Network*, Bellevue Literary Press.

Mori, M. (2012) "The Uncanny Valley" (translated by K. F. MacDorman & N. Kageki), IEEE Spectrum (originally published in Japanese 1970), https://spectrum.ieee.org/the-uncanny-valley#_ftn1

Nussbaum, M. C. (2001a) "Emotions and human societies", in *Upheavals of Thought: The Intelligence of Emotions* (pp. 139–173), Cambridge University Press.

Nussbaum, M. C. (2001b) "Introduction", in *Upheavals of Thought: The Intelligence of Emotions* (pp. 1–16), Cambridge University Press.

Nussbaum, M. C. (2013) *Political Emotions: Why Love Matters for Justice*, Harvard University Press.

O'Leary, L. (2022, January 3) "How IBM's Watson Went from the Future of Health Care to Sold Off for Parts", *Slate*, https://slate.com/technology/2022/01/ibm-watson-health-failure-artificial-intelligence.html

OECD. (2022) *The Culture Fix: Creative People, Places and Industries, Local Economic and Employment Development (LEED)*, OECD Publishing, https://doi.org/10.1787/991bb520-en

OECD (2024, July 20) *OECD.AI Policy Observatory*, https://oecd.ai/en/dashboards/overview

Onfray, M. (2015) *A Hedonist Manifesto the Power to Exist* (p. 104), Colombia University Press.

Pasquale, F. A. (2020) *New Laws of Robotics: Defending Human Expertise in the Age of AI*, Harvard University Press.

Pasquale, F. A. (2015) *The Black Box Society – The Secret Algorithms That Control Money and Information*, Harvard University Press.

Penta, L. J. (1996) "Hannah Arendt: on power", *The Journal of Speculative Philosophy, New Series*, 10(3), 210–229.

Picard, R (2017, June) "Affective computing, emotion, privacy, and health, artificial intelligence podcast by Lex Fridman", https://www.media.mit.edu/articles/rosalind-picard-affective-computing-emotion-privacy-and-health-artificial-intelligence-podcast/

Picard, R. W. (2000) *Affective Computing*, First, Massachusetts Institute of Technology (originally published in 1997).

Podoletz, L. (2023) "We have to talk about emotional AI and crime", AI & Society, 38, 1067–1082.

Qvortrup, J. (n.d.) "Childhood and Politics", https://www.diva-portal.org/smash/get/diva2:1397915/FULLTEXT01.pdf

Rand, E. K. (1932, April) "The humanism of Cicero", *Proceedings of the American Philosophical Society*, 71(4), 207–216.

Rands, W. B. (1922) "The world: a child's song", in Arthur Quiller-Couch (ed.), *The Oxford Book of Victorian Verse*. Oxford.

Rao, L. (2011, July 3) "Sexual activity tracked by fitbit shows up in Google search results", http://techcrunch.com/2011/07/03/sexual-activity-tracked-by-fitbit-shows-up-in-google-search-results/

Richard, B. (2015) *Jonathan Livingston Seagull*, Harper Thorsons (originally published 1970).

Richards, O. (2007, November 28) "World exclusive: the Joker speaks. He's a cold-blooded mass-murdering clown", *Empire*, https://www.empireonline.com/movies/news/world-exclusive-joker-speaks/

Richeson, A. W. (1940, November), "Hypatia of Alexandria", *National Mathematics Magazine*, 15(2), pp. 74–82.

Rizvi, J. (2019, June 30) "Open-plan work spaces lower productivity and employee morale", *Forbes*, https://www.forbes.com/sites/jiawertz/2019/06/30/open-plan-work-spaces-lower-productivity-employee-morale/#46528a461cda

Roose, K. (2023, February 17) "A conversation with Bing's chatbot left me deeply unsettled", *New York Times*, https://www.nytimes.com/2023/02/16/technology/bing-chatbot-microsoft-chatgpt.html

Sabelo Mhlambi quoted in Miller, K. (2022, March 21) "The Movement to Decolonize AI: Centering Dignity Over Dependency", HAI Stanford University.

Sartre, J. P. (1989) "Existentialism is a humanism" (translated by Philip Mairet), in Walter Kaufman (ed.), *Existentialism from Dostoyevsky to Sartre*, Meridian Publishing Company (originally in French 1946). https://www.marxists.org/reference/archive/sartre/works/exist/sartre.htm

Schröder, S. (2021) "Humanism in Europe", in Anthony B. Pinn (ed.), *The Oxford Handbook of Humanism, Oxford Handbooks Series* (2021; online ed., Oxford Academic, 4 Oct. 2019).

Searle, J. R. (1980) "Minds, brains, and programs", *Behavioral and Brain Sciences*, 3(3), 417–457.

Sen, A. (2002, January 5) "How to Judge Globalism", The American Prospect, https://prospect.org/features/judge-globalism/

Shakespeare, W. (1984) *Macbeth* (edited by Kenneth Muir), Methuen Drama (originally published 1623).

Shelley, M. (1989) *Frankenstein*, Puffin Books (originally published 1818).

Shneiderman, B. (2022) *Human-Centered AI*, Oxford University Press.

Siddarth, D., et al. (2021). "How AI Fails Us", *Justice, Health, and Democracy Impact Initiative & Carr Center for Human Rights Policy*, https://ethics.harvard.edu/files/center-for-ethics/files/aifailsus.jhdcarr_final_2.pdf?m=1651510742

Šimić, G., et al. (2021) "Understanding emotions: Origins and roles of the amygdala", *Biomolecules*, 11(6), 823.

Simmel, G. (1909, November), "The problem of sociology Georg Simmel", *American Journal of Sociology*, 15(3), 289–320, The University of Chicago Press.

Simmel, G. (2010) *The View of Life: Four Metaphysical Essays with Journal Aphorisms* (translated by J. A. Y. Andrews & D. N. Levine, from original German 1918), University of Chicago Press.

Simonite, T. (2020, October 26) "How an algorithm blocked kidney transplants to black patients", *Wired*, https://www.wired.com/story/how-algorithm-blocked-kidney-transplants-black-patients/

Sinclair, M. (2019) "Bergson's philosophy of art" in A. Lefebvre and N. F. Schott (eds.), *Interpreting Bergson*, Cambridge University Press.

Smidi, A., Shahin, S. (2017) "Social media and social mobilisation in the Middle East: A survey of research on the Arab Spring", *India Quarterly*, 73(2), 196–209.

Smuha, N. A. (2022) "The Human Condition in an Algorithmized World: A Critique through the Lens of 20th-Century Jewish Thinkers and the Concepts of Rationality, Alterity and History," Institute of Philosophy, KU Leuven.

Spence, J. (2003) *Becoming Jane Austen*, Hambledon Continuum.

Stoycheff, E. (2016, March) "Under surveillance: Examining Facebook's spiral of silence effects in the wake of NSA internet monitoring", *Journalism & Mass Communication Quarterly*, 93(2), 296–311.

Suda On Line. (n.d.) *Upokrisis (upsilon 166)*. http://www.stoa.org/sol-entries/upsilon/166

Talaat, F. M. (2023) "Real-time facial emotion recognition system among children with autism based on deep learning and IoT", *Neural Computer & Application*, 35, 12717–12728.

Talbot, M. (2016, July 17) "The History of Crowd Control and the Cleveland Convention", *The New Yorker*.

Tarr, J. A., McShane, C. (2008) "The horse as an urban technology", *Journal of Urban Technology*, 15(1), 5–17.

Thoreau, H. D. (1849) *On the duty of civil disobedience*, https://www.gutenberg.org/files/71/71-h/71-h.htm

Thorpe, B. (1907) *The Elder Edda of Saemund Sigfusson, and the Younger Edda of Snorre Sturleson*. Norroena Society.

Thuma, A. (2011) "Hannah Arendt, agency, and the public space", in M. Behrensen, L. Lee & A. S. Tekelioglu (ed.), *Modernities Revisited, IWM Junior Visiting Fellows' Conferences Proceedings*, Vol. 29.

Tomkins, S. S. (1962) *Affect, Imagery, Consciousness: The Positive Affects* (Vol. 1), Springer Publishing Company.

Turing, A. (2004) "Computing machinery and intelligence", in J. B. Copeland (ed.) *The Essential Turing: The Ideas That Gave Birth to the Computer Age* (originally published in 1950), Clarendon Press.

Ucnik, L. (2022) "Hannah Arendt's action and contemplation: Two sides of the Same coin", *Journal of Social Philosophy*, 53(1), 76–92.

UN. (2013) 68/167 "The right to privacy in the digital age", *Resolution adopted by the General Assembly on 18 December 2013*.

Vaughan, M. (2007) "Introduction: Henri Bergson's Creative evolution, *Substance*, 36(3, 114), in *Henri Bergson's "Creative Evolution" 100 Years Later* (pp. 7–24), The Johns Hopkins University Press.

Vincent, J. (2016, March 24) "Twitter taught Microsoft's AI chatbot to be a racist asshole in less than a day", *The Verge*, https://www.theverge.com/2016/3/24/11297050/tay-microsoft-chatbot-racist

Virilio, P. (1989) *War and Cinema: The Logistics of Perception*, Verso.

Whaanga, H. (2020) "AI: A New (R)evolution or the new colonizer for Indigenous peoples?" in J. Lewis (ed.), *Indigenous Protocol and Artificial Intelligence Position Paper*, The Initiative for Indigenous Futures and the Canadian Institute for Advanced Research (CIFAR).

Weber, M. (2001) *The Protestant Ethic and the Spirit of Capitalism* (Introduction by A. Giddens), Routledge (original translation from German by T. Parsons 1930, Original German text 1905).

Weil, S. (1970) *First and Last Notebooks* (translated by R. Rees) (original work published 1950), Oxford University Press.

White, T. (n.d.) "What Hannah Arendt really mean by the banality of evil?", *Aeon*, https://aeon.co/ideas/what-did-hannah-arendt-really-mean-by-the-banality-of-evil

Wiener, N. (2013) *Cybernetics or, Control and Communication in the Animal and the Machine*, 2nd ed., Martino Publishing (originally published 1948).

Wiener, N. (2013) *Cybernetics or, Control and Communication in the Animal and the Machine*, 2nd ed. (originally published in 1948), Martino Publishing.

Williams, A. (2024, February 14) "When love and the algorithm don't mix", https://time.com/6694129/dating-app-inequality-essay/

Williams, R. (2021, July 15) "I did nothing wrong. I was arrested anyway", https://www.aclu.org/news/privacy-technology/i-did-nothing-wrong-i-was-arrested-anyway

Williams, R. (1993) "Culture is ordinary" in A. Gray & J. McGuigan (eds.), *Studying Culture: An Introductory Reader*, Edward Arnold (originally published 1958).

Woolgar, S. (1987) Reconstructing man and machine: a note on sociological critiques of cognitivism", in W. E. Bijker, T. P. Hughes, & T. Pinch (eds.), *The Social Construction of Technological Systems* (pp. 311–328) MIT Press.

World Economic Forum. (2023) *Davos23: Five Tech Superpowers Transforming Humanity and Technology*. World Economic Forum Agenda, https://www.weforum.org/agenda/2023/01/davos23-five-tech-superpowers-humanity-technology/

World Health Organization. (2021) *Ethics and Governance of Artificial Intelligence for Health: Who Guidance*, https://iris.who.int/bitstream/handle/10665/341996/9789240029200-eng.pdf?sequence=1

Wynsberghe, A., Robbins, S. (2022) "Our new artificial intelligence infrastructure: becoming locked into an unsustainable future", *Sustainability*, 14(8), 4829.

Yang, Z. (2022, November 22) "China just announced a new social credit law. Here's what it means", *MIT Technology Review*, https://www.technologyreview.com/2022/11/22/1063605/china-announced-a-new-social-credit-law-what-does-it-mean/

Žižek, S. (2008) *The Fragile Absolute: Or, Why Is the Christian Legacy Worth Fighting For?*, Verso.

Index

Note: Page numbers followed by "n" refer to notes.